THE LIFE OF
BIRDS

THE LIFE OF
BIRDS

David Attenborough

First published in 1998 by BBC Books,
an imprint of BBC Worldwide Ltd.,
80 Wood Lane, London W12 0TT

Frontispiece: an Egyptian vulture soaring, Oman

ISBN 0563 38792 0

Printed in Great Britain by Butler and Tanner Ltd., Frome

CONTENTS

FOREWORD

It is easy to understand why so many of us are so fond of birds. They are lively; they are lovely; and they are everywhere. They have characters with which we can easily identify – cheeky and shy, gentle and vicious, faithful – and faithless. Many enact the dramas of their lives in full view for all to see. They are part of our world yet, at a clap of our hands, they lift into the air and vanish into their own with a facility that we can only envy. And they are an ever-present link with the natural world that lies beyond our brick walls. It is hardly surprising that human beings have studied birds with a greater dedication and intensity than they have lavished on any other group of animal.

The first task of ornithology was to give names to birds. Every society, of course, has produced its own version, often in great detail. In the eighteenth century, the Swedish naturalist, Carl von Linné, proposed a uniform way of classifying all living things based on the names used by Greek and Roman naturalists. That, greatly refined and elaborated, remains the system used by scientists all over the world ever since. Today, two hundred and fifty years later, we have found names for some ten thousand different species of birds. Museums and other scientific institutions have accumulated cabinets full of bird specimens with dozens, sometimes hundreds of examples of each species, each one carefully prepared, meticulously measured with every tiny variation in coloration and size duly noted. The introduction of portable binoculars and, later, photography, allowed that high expertise to spread into the field. Now it is no longer necessary to shoot a bird to identify it. Now ornithologists have become so expert that they can identify a wild living bird from a snatch of song or the briefest glimpse of its plumage or silhouette. That is a skill which I greatly admire, but one, alas, that I do not possess.

But that is not what this book is about. My fascination with birds comes from watching how they behave. Ornithologists began to study this aspect of their subjects rather earlier than those working in many other branches of zoology. While big-game hunters were still shooting antelopes in the belief that establishing the maximum size of the horns of any species told us something important and were arguing, on the basis of skin patterns, how many species of giraffe exist, ornithologists were beginning to investigate the journeys birds make. To do this, they needed to identify individuals. One technique, which those working with other animals were somewhat slower to adopt, was to use tags or bands. Jean-Jacques Audubon, back in the 1820's, tied coloured threads to the legs of the flycatchers that each summer visited his parents' mill in Pennsylvania and so established that birds which nested there reappeared the following spring after their migration south and nested in exactly the same place. That in itself was an astonishing finding, but a mere hint of what would be discovered about the extraordinary abilities of birds.

Those discoveries, however, were a long time coming. Individual birds of the same sex and species tend to resemble one another more closely than do those of any other large animal. Scientists working with elephants quickly learn to distinguish individuals from the irregular shapes of their huge ears. Chimpanzees have faces that are as different from one another as those of human beings. Humpback whales have different white patterns on the flukes of their black tails. Lions, being of a quarrelsome disposition, tend to acquire characteristic scars on their muzzles – and even if they don't, have whiskers that vary in their number and disposition. But it is virtually impossible to distinguish one fully fit male hedge sparrow from another.

So watching one in our garden, we tend unthinkingly to make assumptions about its social arrangements. We assume that it is always the same one which collects a worm from the lawn and seldom question whether it is also always the same bird which devotedly supplies food for the chicks in the nest in the hedge. Until comparatively recently, it simply did not occur to anyone that there was any need to fit a ring on a bird that lived permanently in the garden to test such assumptions. But when an ornithologist caught hedge sparrows and did so, he discovered that the marital arrangements of these birds were of such a variety that had they been human beings they would have made headlines in the newspapers.

Once the use of leg-rings extended beyond migration studies, other techniques soon came into use. Small radio transmitters were fitted on the backs of large birds which sent regular signals to a satellite miles high in the sky and relayed them down to a receiving station on earth. So it was discovered that a wandering albatross may travel a thousand miles in order to gather a cropful of food and bring it back to its chick sitting by itself on a lonely Antarctic island. Penguins were persuaded to swallow tiny instruments that measure water pressure. These, carried in their stomachs, showed that king penguins regularly swim down to depths of a thousand feet in the ocean in search of fish. Genetic fingerprinting was used to identify the exact parentage of young birds, and it was revealed that such was the complexity of the Australian fairy wren's life that a male may not be the father of a single one of the chicks that he so devotedly feeds in his nest. The science of bird behaviour has now become a rich and fascinating subject.

And that is the subject of this book. Even though many of the investigative techniques are of such sophistication that they can only be used by full-time research scientists, there is still a great deal of research work that is done by dedicated amateurs. The ubiquity of birds and the devotion they inspire has produced a worldwide army of enthusiasts prepared to devote their spare time and endure the most uncomfortable of conditions in order to make observations and collect data. The literature recording all this work, both amateur and professional, is huge and widely scattered. Much of it is in specialised journals only available in a scientific library. Without that great body of raw data and the distillations that have been made from it, barely a single page of what follows could have been written. My debt to such publications is huge. I have not, however, given individual references to all these sources in order not to clog the text. Nor, for a similar reason, have I used scientific names for the species I mention. Their precise identity, however, can be discovered by consulting the index in which their scientific name is listed beside the colloquial name I have used in the text.

It is my hope that the pages that follow will show something of the deep fascination to be found not only in naming birds but in discovering what they do and why they do it.

1

TO FLY OR NOT TO FLY

A bird of prey circles high in the sky above a limestone cliff that rises sheer and white above the green forests of Borneo. Earlier in the day, it had been roosting in the trees so motionless that few other animals could have noticed it. As the evening sun started to sink and lose its brilliance, the bird had flapped lazily into the air to start its patrol. Now it looks down towards the black mouth of a cave that gapes in the face of the cliff. Deep inside, a million bats hang from the ceiling, packed together so tightly that the rock above them is hidden. Unlike the hawk, they cannot see the sinking sun to take a cue from its impending disappearance. Nor can they have detected a drop in temperature for the conditions inside the cave are remarkably stable. Yet somehow the bats know that outside the day is coming to an end and that soon they will be able to fly out into the darkening forest to gather their nightly harvest of insects. A few flutter uncertainly into the air and fly to and fro, navigating in the blackness by the echoes of their ultrasonic squeaks. Then, within a minute or so, they organise themselves into a column. Like a wavering black ribbon, it snakes across the ceiling only inches beneath the rock. It advances, winding round bumps and along crannies from one chamber into the next, until eventually it reaches the great hall that forms the cave's entrance. Outside, the sun has vanished but if you are sitting in the cave mouth there is still enough light for you to watch the black ribbon advancing diagonally across the ceiling until it reaches the highest point in the far corner. As it arrives, the bats break ranks and spill out into the open across the forest canopy.

The hawk watches. It does not seem to be in any hurry. There are so many bats in the cave that their exodus will last many minutes. It can choose its moment. Suddenly, it makes up its mind. It tips down, accelerates on rapidly beating wings and dives straight into the cloud of bats. Its feet come forward and it grabs a bat with its talons. Sometimes it rips its prey apart with its beak as it flies. On other occasions, it carries the crumpled bat back to its roost to feed. It may catch two or three bats before the last of them leaves the cave and darkness falls. The hawk has particularly large eyes but even so, after about half an hour, it can no longer see well enough to repeat its accurate high-speed pounce. However, it has gathered all it requires. Bats are certainly the most skilled and agile of all flying mammals but they are no match for the bat hawk. As long as there is light, the skies belong not to the mammals but to birds.

Perhaps that is to be expected. Birds have been flying for much longer than bats. The oldest bat fossils to be discovered date from around fifty million years ago and birds were flying at least a hundred million years earlier still, at the time of the dinosaurs. They were not, however, the first animals to colonise the air. Insects preceded them by some two hundred million years. Some of the first were giants with wings over a foot across. As the millennia passed, the descendants of these pioneer aviators evolved many different techniques for flying. Some had two pairs of wings, others just one. Some developed gyroscopic stability controls, others beat their wings faster than muscles could contract and managed to do so by hitching them to the vibrating shell of their thorax. But they did not grow very large. Insect bodies are constructed in such a way that they cannot operate above a certain size and no insect has ever exceeded those early giants. When birds appeared, the insects found themselves outflown and that superiority remains today.

If you are in doubt, watch a spotted flycatcher, that common summer visitor to gardens all over Europe. It is a somewhat nondescript bird with no dramatic colour in its plumage, but it draws your eye immediately because of its actions. It usually sits, very upright, on a bare branch of a tree and every few seconds takes off in a swerving twisting flight before returning to its branch. Get nearer to it and you may hear a faint click in the middle of its excursions. That is the sound of its beak snapping shut on an insect. Dragonflies may zig-zag and dodge but they are lucky to escape if the bird gets anywhere near them. Flies and ichneumons are snatched from the air without any difficulty. The bird is so skilled that sometimes it returns to its perch with several of small insects held deftly in its beak. Each victim is dealt with appropriately. Butterflies are stripped of their wings. Horse-flies are swallowed as they are, but bees and wasps, although they are of a similar size and general appearance, are recognised and made harmless by being rubbed vigorously against the perch so that their stings are discharged before the bird swallows them. The insects met their match in the air a long time ago.

When, then, did the birds first achieve this dominance of the skies? The answer was discovered during the last century in Bavaria not far from Munich. The countryside there is studded with quarries where men extract a lovely cream-coloured limestone that has been used for building since Roman times. It is so smooth,

uniform and fine-grained that, in the nineteenth century, it was used for lithographic printing. In some places, it splits into thin slabs. Separating them, one after the other, from the top of a block is like opening the pages of a book. Most of these pages, it must be said, are blank, but every now and then, lifting one reveals the perfectly printed record of an animal – a bewhiskered shrimp, a fish with its fins and ribs immaculately preserved, a horseshoe crab lying at the end of a trail of its last footprints. Sometimes there is a faint brown stain around these remains showing the position of soft parts that decayed and dissolved soon after death. Even such insubstantial creatures as jellyfish have been found delicately delineated.

From these fossils it is not difficult to deduce the sort of conditions under which the limestone was deposited. It was once mud on the floor of a warm tropical lagoon. Land lay a few miles away to the north and to the south a coral reef separated the lagoon from the open sea. Because there was little flow in or out and the rate of evaporation in the warm sunshine was high, the waters became so saline that no animal could make the lagoon its permanent home. But periodically an unusually high tide would sweep over the reef, carrying with it creatures from the sea beyond.

Above: a spotted flycatcher seizing
a butterfly, Europe

Occasionally animals from the mainland flew across the lagoon and crashed into its tepid salty waters. So there are fossils of insects – mayflies, dragonflies, locusts, beetles, wasps – ample evidence of how advanced and sophisticated insects had become in the air even at this remote period. There are also exquisitely preserved skeletons of small flying reptiles – pterosaurs, with the faint outlines of their skinny wings plainly visible around their elongated straw-thin fingers.

Such wonderful, beautiful fossils have been collected and treasured for centuries. But in 1860 a quarryman working near the small village of Solnhofen split a block and made an unprecedented and astonishing discovery – a feather. It is quite small, only six inches in length and less than two and a half inches wide, but it is preserved in the greatest detail. In a couple of places the filaments of the blade are separated and at the base of the quill there is a little tuft of isolated ones. The vane on one side of the quill is only half the width that it is on the other. This asymmetry has particular significance. The feathers on a modern bird's wing are shaped in just this way and positioned on the wing with the narrow stronger edge at the front. Such a shape makes it clear that this ancient feather had an aerodynamic function. It looks scarcely different from a wing feather that one might pick up today on a country walk. Yet it must have fallen from the wing of a creature that flew across the lagoon one day a hundred and fifty million years ago.

What kind of creature was it? No living creatures grow feathers except birds. Indeed, feathers are taken to be the defining characteristic of birds, so the Solnhofen feather, by definition belonged to a bird. But what kind of bird? Science did not have to wait long for the answer. The very next year, in 1861, in a quarry only a few miles from that where the feather had been found, a nearly complete skeleton was discovered. It was the size of a chicken and it was surrounded by the detailed impressions of feathers.

But this was a very strange bird indeed. It had a long bony tail; each fore-leg had three separate digits, each of which ended with a curved claw; and its skull, as was revealed by later finds, carried not a beak but bony jaws studded with teeth. It was clearly part-reptile and part-bird. The scientist who described it called it Archaeopteryx, a name based on two Greek words meaning 'ancient wing'. Since that time seven more specimens have been identified so that now we know a great deal about this extraordinary creature's anatomy. Yet the debate still continues about exactly how it lived.

The claws on its wing fingers give some clues. A few birds still retain such things even today. Some swans, ducks, jacanas and several other birds have them, hidden out of sight beneath their plumage. The screamer, a goose-like bird that lives in South America, carries two very prominent and easily visible ones on each wing which it displays during territorial disputes and even uses as weapons during fights between males. But perhaps the most likely clue as to how Archaeopteryx used its

Archaeopteryx from Solnhofen, Germany

wing claws comes from an odd bird living in Venezuela and Guyana, called the hoatzin.

The hoatzin is a swamp-living leaf-eater with a rather clumsy lumbering flight. It makes an untidy platform of twigs as a nest on which it lays two or three eggs. The young, when they hatch, have two well-developed claws on each wing. As the nestlings grow, they become venturesome and clamber about in the mangrove trees using these wing claws to help them cling to the branches. If they become alarmed, perhaps by some predator, they will dive into the water beneath and then, after a short time, use their claws to clamber back to their nest. Once the birds become adult, they lose their claws. Maybe Archaeopteryx which retained them throughout its life used them in the same way as the hoatzin chicks do. Although no fossil logs or substantial branches have been found in the Solnhofen limestone, there are occasional leaves of conifers, cycads and maidenhair trees so undoubtedly forests were not far away. The rarity of the Archaeopteryx fossils suggests that these creatures only appeared above the lagoon very infrequently. They must have strayed across from the mainland and the forests that were their true home.

There are other indications that Archaeopteryx lived in trees. The big toe on each foot points backwards, as it does in most modern birds, so enabling the animal to grasp a perch. Its long tail also seems to belong to a bird that lived up in trees. It is beautifully preserved in several of the specimens and shows no sign of being bedraggled at the tips as it might well be if the creature spent much time on the ground. And it is so long that it would be a grave encumbrance if its owner lived on water.

But did it use its wings merely to glide from one branch to a lower one, or was it capable of beating them and so powering its flight? If it could flap, then it must have had muscles connecting its wings to the bones of its chest. There was no evidence in the first specimens of any bone that might have provided such an attachment, but this does not mean that such muscles did not exist. They could have been fixed to a piece of cartilage which would have rotted and left no trace to be fossilised. So the question had to remain open. But in 1992 new evidence appeared. Yet one more specimen, the seventh, was found in a quarry only a few miles from that which produced the first. It is smaller than the first and differs sufficiently in other details for it to be regarded as a different species. It has been named Archaeopteryx bavarica and it has something the other specimens lack, a large bony breastbone. This is more than adequate as an anchor for wing muscles. So the likelihood is that these pioneering aviators did not just glide, but flapped their way through the forest – and occasionally across the lagoon into which some of them crashed.

Archaeopteryx could not have been the first backboned animal to have taken to the air. Its feathers have such a complex structure that they must be the product of a long evolutionary process that extended over many thousands of generations. But

Right: a hoatzin nestling, Venezuela

why should that process have started? Why should Archaeopteryx's ancestors have found it advantageous to have feathers even of the simplest kind? The answer to that clearly depends on who those ancestors were.

One possibility is that they were dinosaurs. Indeed, the similarity between Archaeopteryx and a small dinosaur is so close that one specimen lay for decades in a museum classified as a dinosaur until a more careful inspection revealed the faint impression of feathers around the forelimbs and made it clear that it was, in fact, an Archaeopteryx. Such small dinosaurs were probably active predators that chased after their prey. To be as active as that, an animal's body must be warm so that its chemistry works vigorously and produces a lot of energy. Fast-moving reptiles of today, such as lizards and snakes, achieve this condition by warming themselves in the sun. But some argue that small dinosaurs such as Velociraptor were able to generate their own heat internally, as mammals do. Such a process is very expensive in terms of energy and uses up a significant proportion of the calories taken in as food, but it brings considerable advantages. It would, for example, enable an animal to be active in the early mornings and gather food when its competitors were still cold and torpid. A warm insulating coat would then be invaluable. Reptiles of the time were covered in scales, as modern ones like the Australian shingleback lizard are today. If those scales increased in length and became fibrous then they might well provide such beneficial insulation.

Now imagine such a creature using its abundant energy to pursue its prey, say a large insect. It might well rise on its hind legs, as the frilled lizard of Australia does when it wants to move at speed. That would leave its fore-legs free. If they were covered with long fibrous scales – proto-feathers – then stretching them out might lift the animal into the air and enable it to snatch at its prey with its mouth. Alternatively, if it was running to escape a bigger animal, then such a manoeuvre might take it out of range and into safety. So this warm-blooded reptile would have taken its first step towards flight. Disbelievers in this theory maintain that such an animal would not, when running on its hind legs and seeking to put on a turn of speed, suddenly stick out its forelegs since such an action would instantly increase its drag through the air and thus inevitably slow it down.

Disregarding this objection, the question must still be asked if the dinosaurs that might have been Archaeopteryx's ancestors were, in fact, warm-blooded? Some maintain that they were. They seek evidence from several sources – from the ratio between their numbers and those of the vegetarian grazing reptiles on which they preyed; from the microscopic structure of their bones; and from the size of their brain. But none of this evidence is accepted by everyone as conclusive. Now other research suggests that they, like reptiles alive today, were *not* able to maintain their bodies at a constant temperature. Such creatures grow at different speeds at different times of the year. In consequence, their bones develop concentric rings somewhat similar to the annual rings in a tree trunk. Such rings have been found in the bones of fossil birds that succeeded Archaeopteryx in the ancient skies. This suggests that Archaeopteryx itself did not generate its own body heat either. Archaeopteryx fossils are so rare and precious that no-one has yet sectioned one of the limb bones to confirm that this is so, but if it is, then the theory that feathered flight originated with creatures that ran along the ground is greatly weakened. Clearly, a cold-blooded animal would be very unlikely to develop an insulating coat of fibrous scales since that would prevent or at least hamper its owner from warming itself in the sun.

There is an alternative hypothesis. Perhaps the ancestral reptile, with or without the benefit of internally generated warmth, started to clamber four-leggedly up into the trees. There are several reasons why it might have done so. Maybe it was taking refuge from larger predatory reptiles on the ground; maybe finding a safe place for its eggs; maybe pursuing insects that lived in the branches. Once it was up in a tree it would need to move about. It is much quicker and less energy-consuming to jump from one tree to a lower branch in a neighbouring one than to descend, run across the ground and climb back up again.

Several animals today leap around in trees and have developed structures to enable them to greatly increase the length of their jumps. In the forests of Borneo alone, at least four different techniques have evolved. The giant flying squirrel has a loose-fitting furry skin that is so voluminous that when it gallops up a tree-trunk it appears to be wearing a floppy cloak. Just how big the garment is you can see when it decides to travel to a neighbouring tree. It leaps off a branch into the air, spread-eagles its legs and reveals that the cloak extends to its wrists and ankles so forming a huge rectangular parachute. This is so effective that the animal can easily glide for a hundred yards. What is more, it can steer, by flexing the long furry tail that streams out behind it and by changing the position of its legs to vary the tension in its parachute. So it can land with great accuracy on a chosen spot such as the entrance to its nest hole.

A small tree-living lizard called Draco glides in a different way. Its ribs are hugely elongated. When the animal is sitting quietly on a branch, they lie close to its body, parallel to its spine to which they are hinged. When it jumps, its abdominal muscles contract, drawing its ribs forward so that they fan out and expose a wide flap of brightly coloured skin on either side of the body. These little lizards are so competent in the air that they constantly flit from one branch to another, pouncing on an insect that has incautiously settled within range, confronting a rival that has made a challenge for territory.

Then there is a frog which has webbed toes so elongated that when it launches itself into the air, each foot acts as a parachute. There is even a gliding snake. When it takes to the air, it pulls up its abdomen towards its spine so that its underside

becomes concave, and bends its long body into zigzags so that it forms a squarish rectangle that is surprisingly effective in catching the air.

All these Bornean creatures, and many more in forests elsewhere, make it quite clear that gliding is a valuable talent for tree-living animals and that they readily evolve modifications to their bodies that enable them to do so. What more likely therefore than that Archaeopteryx's ancestors also developed this ability and that they did so, in this case, by modifying their scales into feathers.

Scales and feathers both originate as tiny pouches in the skin. Both are constructed of similar horny material. Even the tiniest extension of feathery scales would bring aerodynamic advantage so it is not difficult to imagine scales on the arms and tail of an ancient reptile becoming increasingly elongated to extend the animal's leaping range. Gradually the structures improved. A central strengthening quill appeared. The filaments on either side, the barbs, developed tiny hooks that linked them all together to form a single, air-catching surface which, if split, could be repaired by zipping the two together with a beak.

For a century after the initial discovery of Archaeopteryx, the Solnhofen limestone remained the only source of evidence about the very early history of birds. Then in 1995 extraordinary fossils began to appear from China. They came from Liaoning Province in the north east of the country. The mudstones in which they had been found had formed at the bottom of a large lake about 120 million years ago, some thirty million years after the era of Archaeopteryx. They included a great range of fish, frogs, crocodiles and small primitive mammals. Among the reptiles was one, a small dinosaur, called Sinosauropteryx or 'Chinese dragon-bird', which had on its head, neck and along its spine what appear to be a line of downy filaments. These have been interpreted as being primitive feathers, though this view is not yet universally accepted. They do not, however, have either the asymmetry or the interlocking structures that would give them an aerodynamic function. Whether they were thick and abundant enough to serve as insulation for a warm-blooded creature, or whether Sinosauropteryx used them for display is still unclear. The animal may, perhaps, be taken as a direct descendant of one of the very earliest reptiles to have developed feathery scales, that may have evolved before Archaeopteryx and somehow outlived it.

Another species from these deposits, however, was most certainly even closer to a true bird than Archaeopteryx. It was about the size of a magpie. Its body was clothed in feathers and it had three clawed digits on each fore-limb. But the column of bones in the tail is much shorter and, most significantly of all, instead of jawbones and teeth it has a horny beak. This must have been a most significant advance towards efficient flight for the change not only reduced the animal's weight but brought its centre of gravity into its abdomen so that it would have no longer been nose-heavy in the air. It was called, in honour of the great Chinese sage,

Left: a flying lizard, Draco, Borneo

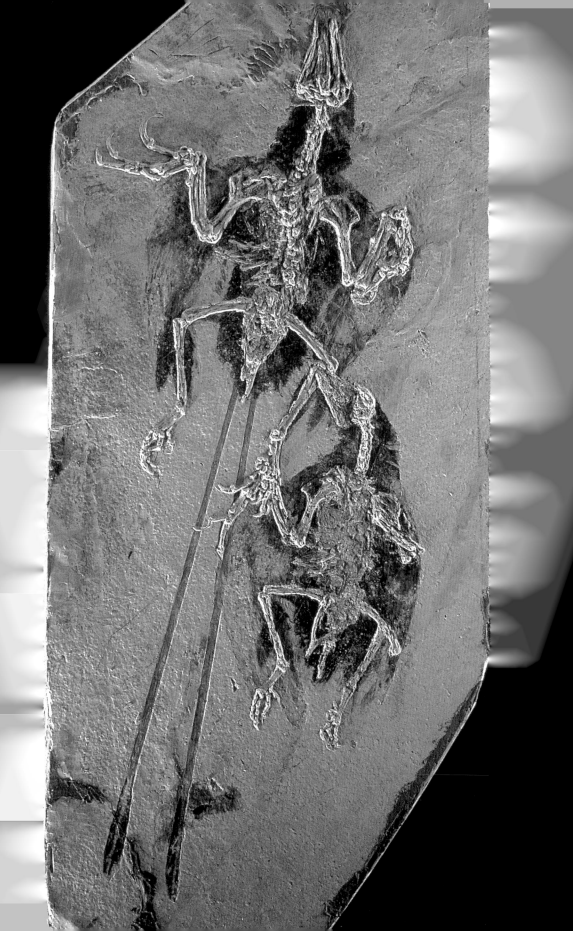

Confuciusornis. Several hundred specimens of this fascinating creature have now been found, some remarkably well preserved with a blackish haze around the bones showing the extent of the animal's feathered flesh. Some even have a pair of extremely long quills projecting from the tail, suggesting that these were males and that there were now major differences between the sexes. The specimens occur in such numbers that it is tempting to suppose that the birds lived in dense colonies.

More recent rocks in Australia and Spain, Argentina and North America have now yielded bird fossils of a still more advanced kind. These later species now had all the weight-saving adaptations that are characteristic of true birds. The long bony tail had been reduced to no more than a small triangle of fused bones at the back of the pelvis. The long bones of their wings and legs had become hollow, supported internally by criss-crossing struts. Many had a keel on their breastbone to which large wing muscles were attached. Their ribs had flanges that bound them together and gave their chest strength. The great tribe of birds was clearly, by now, well established.

Then, sixty-five million years ago, a still mysterious catastrophe overwhelmed life on earth. It seems to have been associated with some body from outer space, an asteroid or a comet, which collided with the earth and created such a titanic explosion that debris blocked out the light of the sun and the earth, perhaps for months, was blanketed by continuous darkness. Whatever the cause, the last of the dinosaurs, whose numbers had in any case been diminishing for some time, vanished. So did many other great reptiles. Mosasaurs and plesiosaurs disappeared from the seas and pterosaurs from the skies. And so did a high proportion of the birds. A few groups, however, survived, among them the ancestors of ducks and geese, loons, gulls and shore-birds such as plovers and waders. They now faced great opportunities. Huge areas of the earth that had only recently supported vast and varied animal populations were almost empty and ready for re-occupation. In particular, the skies were no longer dominated by flying reptiles. The ancestors of modern birds evolved to colonise them and they did so at an astonishing speed. Within ten million years all the orders of the birds of today, except small perching birds, had appeared.

One remarkable site on the east coast of Britain at Walton-on-the Naze in Essex has provided astonishing evidence of this swift and rich development. Here the London Clay, which was deposited some fifty million years ago, has yielded over six hundred specimens of ancient extinct birds. Among them are species that can be linked with today's cuckoos, parrots, owls, swifts, hawks, gulls, kingfishers and many more groups. Similar evolutionary explosions were taking place all over the world. In North America great colonies of waterbirds that appear to be ancestors of today's ducks and geese, nested along the shores of a large lake that covered a great part of what is now the state of Wyoming. Their bones and fragments of their

*Left: Confuciusornis fossils, one with a
pair of long tail feathers, from
Liaoning Province, China,*

eggs protrude in thousands from the deposits of sand that were laid down at that time and have now solidified into sandstones.

Not far from this site are deposits that were formed in open grasslands and from these have come the remains of birds that exploited their new opportunities in a radically different way. The dinosaurs that had made the land such a dangerous place for the ancestral birds had gone. The birds could, therefore, once again become terrestrial – and they did. One kind stood over six feet tall and had a huge head a foot and a half long with an immense heavy bill. Its wing bones are tiny and make it quite clear that the bird could not fly. Ever since its discovery over a century ago, this bird, named Diatryma, has been thought of as a fearsome predator that lived on smaller reptiles and other little creatures. Recently, some scientists have pointed out that its beak does not have a hooked tip and have suggested that maybe Diatryma was not an active hunter but fed instead on carrion or even leaves. But even if that were the case, it was not the only huge flightless bird stalking across the grasslands of the world. There were others in South America and these, certainly, were hunters. Some stood over nine feet high and had heads as big as those of horses and beaks armed with formidable hooks. They have been called, not without reason, terror-birds.

But the birds' bid for dominance of the land as well as the skies did not go unchallenged. Ever since the time of Archaeopteryx, small shrew-like animals had been running around on the floor of forests searching for insects. They had already acquired warm bodies which were insulated with hair. They lived in a modest way right through the age of the dinosaurs. They too, like the birds, survived the great extinction and now they too had a chance. They increased in size. Some, judging from their teeth, became ferocious and not dissimilar from today's hyenas. It was with such creatures, the early mammals, that the flightless birds had to contest the dominance of the land. The mammals evolved into larger and more varied forms and, ultimately, they won. Twenty-five million years ago, Diatryma and the terror birds had all disappeared.

Giant flightless birds, however, do still stalk the earth – the ostrich in Africa, the rheas in South America, and the emu and the cassowaries in Australia and New Guinea. It is tempting to suppose that these too appeared, together with Diatryma and the terror birds, soon after the great extinctions. At that time, the continents of the world were not in the positions they are today. 150 million years ago, they had all been part of one super-continent. That over a very long period broke up and the fragments drifted apart to become isolated as today's continents. Perhaps the ancestors of today's flightless birds travelled with them and, over the millions of years since then, evolved into the separate species on different continents that we know today.

Right: a cassowary with eggs,
northern Australia

The alternative explanation is that ostrich, rheas, emu and cassowaries evolved separately from flying forebears and that the resemblances between them are due to these huge flightless birds all becoming adapted to a similar life-style. It is certain, however, that none of them are immediately descended from reptiles. They are not, as it were, half-way stages to full flight. Many characteristics of their bodies indicate very clearly that they are descended from flying ancestors. The bones of their stunted wings are fused together, just as they are in the wings of a flying bird. The feathers on these wings are placed in just the same way as true flight feathers – a disposition that makes no sense if the birds had developed these feathers just to keep themselves warm. They also have the weight-saving beak, and the little triangle of bone at the rear that is all that remains of the reptilian tail.

Their feathers however have become degenerate. The hooks on the barbs which joined them together into an aerodynamic vane have disappeared. In the ostrich, the feathers have become soft and fluffy, and in the rheas, which live in a colder part of the world, long and shaggy. These modern flightless birds have survived in competition with mammals because their stature and their long legs enable them to run at great speed and give them a real chance of outpacing any mammal that might seek to attack them. It is not insignificant that the ostrich, rheas and emu live on open grasslands or in semi-deserts where the country is open enough for speed

to be an effective defence. The cassowary has a much stockier build but is essentially a forest-living bird, and it has no need of long legs to give it great speed for there have never been large mammalian predators in the tropical forests of northern Australia and New Guinea where it lives.

Birds have repeatedly abandoned flight during their history. Flying is very expensive in terms of energy and birds do not travel by air if it is safe for them to do so by land. The change can be very swift. The Galapagos Islands, lying seven hundred miles west of the coast of Ecuador out in the Pacific, rose from submarine volcanoes not more than five million years ago. At various times since, birds from the mainland have managed to reach them, perhaps swept there by freak storms. Among them were cormorants. Once arrived, there was little need for the cormorants to fly. They could find all the food they needed simply by diving into the nearby sea. No mammals had reached the islands that might harry them. Nor did the ancestral cormorants need their wings to propel themselves underwater for all cormorants do that with their feet. As one generation succeeded another without ever flying, their wings dwindled. Today, the wings they hold out to dry after their fishing sessions can be seen to be stunted with tattered feathers that certainly would not sustain them in the air if they were reckless enough to try to fly.

The island of New Caledonia in the western Pacific is home to another flightless bird, the kagu. It is the size of a domestic hen with the general shape of a stockily built heron. Its plumage is pale dove-grey. And that is odd, for this is a forest-living bird that spends its life on the ground searching for insects and other small prey. Most birds elsewhere with such a lifestyle would be camouflaged. The pale kagu is immediately visible – an easy target for ground-living hunters. But there are none. Its wings are still fully-feathered and appear at first sight to be quite adequate to support it in the air. Its muscles, however, are now not large enough to beat them. The greatest use it makes of its wings, aerodynamically, is to help it keep its balance as it moves around on the ground. If it is descending a slope it may even open them and glide a few yards. It also uses them when it displays, either to drive off rivals or to impress mates. Then, when it opens them, it reveals that it is not, after all, a plain pale bird. Its primary wing feathers, which are normally hidden, are strikingly barred with grey. How long ago it took advantage of its immunity from predators and abandoned flight we do not know. The most ancient remains of it to be found on the island are only around two thousand years old. Maybe its wings are not yet as reduced as those of the flightless cormorants on the Galapagos because its isolation on New Caledonia has been even shorter.

Rails, as a family, seem to be particularly prone to abandoning flight. They are small relatives of the moorhen which live in dense vegetation around swamps and

ponds and are widely distributed throughout the world. Their ancestors must have been capable of crossing the sea for they are found on many small islands and there, in isolation, each population has evolved its own characteristics that make it just slightly different from any other elsewhere. And over a quarter of the species that have done that have become flightless. Both Gough Island and nearby Tristan da Cunha in the South Atlantic have their own flightless rail. So, once, did Laysan in the Pacific. All three are small islands built by volcanoes erupting on the sea floor in comparatively recent times, like the Galapagos. Another member of the family reached New Zealand. It too lost its ability to fly and it still seems to have little fear of other ground-living creatures, even though mammals now live alongside it. The New Zealanders call it the weka, and one will appear on the beach of many off-shore islands just as soon as your boat lands, clearly expecting that you, in one way or another, will be a source of food and not hesitating to demand it.

New Zealand is very different from the other comparatively recent islands where flightless rails are found. It is an ancient fragment of land, a piece of the ancient supercontinent that became separated from Australia and Antarctica at the critical time when the dinosaurs had disappeared and the mammals had not yet risen to power. It was in these islands that modern birds, from an early time in their history, eventually exploited flightlessness to the full.

Above: wekas, New Zealand

Among them were the moas, a family, now extinct, that included the biggest bird that has ever existed. They looked somewhat like the ostrich and it is tempting to think that they were related to it. If that were so, then moas must have been wandering about New Zealand when it was still part of the supercontinent and stayed on board as passengers when the island split away and started its long voyage south. The alternative explanation is that the ancestors of the moas had functional wings and flew to the islands very soon after New Zealand's isolation and that, in the absence of mammalian or reptilian competitors on the ground, they took to the flightless way of life and evolved into forms very similar to an ostrich because they lived in a very similar way. So far no fossil evidence has been found that would conclusively settle the question one way or the other.

We do know however that a thousand or so years ago there were as many as eleven different species. In one way they were the most thoroughly flightless birds of all, for whereas ostrich and emu and other flightless birds do have the relics of wings, the moas had lost them entirely. They did not even have the remnants of wing bones within their bodies. The biggest of them, which weighed around 500 pounds or more, was once celebrated as the tallest bird that ever existed. That claim, however, is not as secure as it might seem. In the past many of the skeletons put together by museums and, doubtless, by collectors who wished to sell record-breaking finds to museums, contained vertebrae from several individuals so that the reassembled skeleton was considerably taller than any moa had been in life. But the claim is also based on a misunderstanding of the animal's posture. The giant moas were forest-living creatures as cassowaries are today in the rain forests of tropical Australia and New Guinea. These birds, when they move through the forest, carry their necks more horizontally than vertically, stretched out ahead of them. The facets on the sides of the moas' vertebrae confirm that they too had this posture. The opening at the rear of the skull, where the spinal cord enters the head and joins the brain, indicates the same thing. In a bird with a vertical neck, like an ostrich, this opening is on the underside of the skull. In forest-living moas, it lies at the rear. Nonetheless, it is probably true to say that the giant moa could, if it wished, reach higher than any other bird that has ever lived. It was clearly an immense and formidable creature.

Why should this moa have grown so extremely large? It was probably due to its diet. All the moas were vegetarian. The forest moa, as we know from preserved stomach contents, had a very fibrous diet – twigs and coarse leaves from native forest trees. Such food takes a great deal of digesting and in consequence must be left to stew for a considerable time. For that to happen, a bird needs a particularly capacious stomach and the bigger the stomach, the bigger the body needed to carry it.

The smallest of the family, that lived on the high moorlands, weighed only a tenth as much as the giant forest moa and was scarcely bigger than a large turkey. It presumably had to endure considerable cold up in the mountains for its shaggy hair-like feathers extended right down its legs to its feet. There were medium-sized species too, that stood about four to five feet high. That is about the same size as modern cassowaries and from them one can get an idea of how formidable the moas must have been. During the breeding season, cassowaries hiss and growl and when rival males fight, they even produce extraordinary roars. If you alarm one – and you may do so even before you see it – it will crash away through the forest with such vigour that its feet thud like drum beats. If it happens to have chicks with it, it may not retreat and then it will be time for you to do so. An angry cassowary will attack and strike out with its feet which are so strong that one blow from them can easily rip open a man's stomach.

Since there were no lions, tigers, bears or any other mammalian predators in New Zealand, one might suppose that nothing could attack moas. But birds them-selves can become hunters and the biggest eagle the world has seen, Harpagornis, evolved here. It had a nine foot wingspan, weighed about 30 pounds and had talons as big as those of a tiger. It may have been able to tackle the biggest of the moas. Certainly several skeletons of medium-sized moas have holes pierced in their pelvic bones that exactly match the positions and size of Harpagornis' talons. To cause such injuries, the claws must have penetrated at least two inches of the moa's flesh. One can picture the huge eagle clinging to the back of the terrified moa as it plunged through the forest trying to dislodge its attacker.

The greatest enemies of the moas, however, and the ones that brought about their total extinction within a few centuries, were those they had never faced before – mammals. The first to reach the islands – apart from bats which were able to fly there – were human beings, Polynesians from the more northerly tropical part of the Pacific. For them, the moas represented a uniquely rich source of meat and they hunted them with vigour. As the human population grew, so the forests in which the moas lived began to disappear. Three hundred years ago they were all virtually extinct.

One small and somewhat similar bird however survived – the kiwi. It was once thought to be a small version of the moa but now it is regarded as being more closely related to the Australian emu. If moas can be considered the avian equiva-lents, in mammal-less New Zealand, of deer, then the kiwi must be counted as comparable to a badger. It lives in burrows; it is almost entirely nocturnal; it eats a variety of things, animal as well as vegetable, but it has a particular taste for earth-worms and other small creatures of the soil; and it marks its territory with sign-posts of exceptionally smelly droppings. All these are characteristics of a badger. Even the kiwi's feathers have become transformed into an approximation of coarse fur. The barbs have lost their hooks and each feather, though fluffy at the base, be-comes towards the top, hard, hair-like and waterproof. Furthermore, a few around

Right: a brown kiwi leaving its burrow, New Zealand

the base of its bill have become bristly and enable the animal to feel its way around in the dark. They are the equivalent of a badger's whiskers.

Kiwis are not easy to observe in the wild. They are increasingly rare and with their acute hearing they are able to stalk away through the dark forest before you can get close enough to see them properly. But on Stewart Island, the most southerly of all New Zealand's offshore islands, you stand a good chance of seeing them for they come out of the bush to feed on the sea shore.

It is best to get down to the beach early so that you are there, settled and quiet, before one arrives. As darkness falls, you may perhaps hear an unearthly shriek, repeated a dozen or so times, coming from within the trees that grow down right to the edge of the sand. A male kiwi is proclaiming his ownership of the territory. After a while, in the gloom, you may just detect a small hunched figure materialising at the edge of the bush. It moves silently and slowly farther down the beach, farther into the open. Its shape is hardly bird-like. Its body is almost globular, with no sign of either wings or tail. If hobgoblins existed, this is what they might look like. It stalks cautiously towards the line of kelp left by the sea along the high-tide mark. Once there, it begins to search for sand hoppers and other small creatures, plunging its long bill into the sand right down to its base. Now you can crawl up, moving as quietly as possible to the line of stranded weed and lie down on it, twenty or thirty yards ahead of the kiwi. It comes closer and closer until you can hear, quite clearly, regular short snorting noises. Kiwis have an excellent sense of smell and can detect the presence of little crustaceans buried as much as an inch or so beneath the surface. Their nostrils, uniquely among birds, are placed at the very end of the beak. As the bird plunges it deep into the sand, these nostrils become clogged and because it has a valve within the bill close to its base, it can clear them by blowing down its bill. Closer and closer it gets until – if you are lucky – it is within two feet or so from you. It is so close that you can see its tiny bead-like eyes as it stands motionless, staring at you. But they must be among the most inefficient eyes in the whole bird kingdom. In this dim light they probably can't give their owner an adequate image of anything farther than a few inches away. Then at last it may realise that the lump lying in front of it is not simply a boulder but something more unfamiliar, and with a wide straddling gait, it rushes off back into the bush.

Birds have flown to New Zealand from other continents ever since it attained its separate identity. And they continue to do so even today. The spur-winged plover has only recently arrived from Australia and has now established itself here. The longer ago they arrived, the more different they have become from their colonising ancestors. And several have found that, in the absence of mammals, it was advantageous to give up flying. Elsewhere in the world, the creatures that run along branches collecting buds and nuts are squirrels. There were no squirrels in New Zealand and so the kokako, a member of one of the islands' most ancient lineages,

has taken on the role. It is the size of a pigeon, plain bluish grey with a pair of fleshy wattles at the base of the beak which in the North Island subspecies are cobalt blue and in the South Island one, a deep orange. It can still fly but not strongly and it usually reaches the top of a tree, not by beating its wings, but by flexing its long muscular legs and springing surprising distances from branch to branch as it searches fruit and insects. Only when it gets to the top of the tree is it likely to fly. Then it simply opens its wings and glides down again – just as a flying squirrel does in so many of the world's forests elsewhere.

The weka is not the only member of the rail family to have reached New Zealand and abandoned flight. A kind of coot has also done so and, furthermore, has become a giant. It colonised the moorlands of the high mountain valleys. In general appearance it is very like its European relative, the moorhen, but it is very much bigger and more subtly coloured with deep blue plumage shading to green on the

Above: a kokako,
North Island, New Zealand

back, and a huge brilliantly scarlet bill. Such a hefty meaty bird was greatly relished by Maori hunters, who called it the takahe. It was already rare when Europeans came to the islands. By the middle of the nineteenth century it was regarded as extinct.

Then in 1948, sensationally, a small population was rediscovered living in one of the most remote valleys in the southwest of South Island, a wild unspoilt area known as Fiordland. The signs of the takahe's presence are stalks of tussock grass that have been pulled out from a clump and clipped and nibbled at the base. The bird carefully selects this part of the grass because it is rich in sugars and minerals. But the richness is only comparative. Tussock, at its best, is not very nutritious and the bird has to eat almost continuously throughout the day to get all that it needs. So a takahe leaves behind it a near continuous trail of discarded stems. This, if you are very lucky, may lead you to the bird. Nonetheless, if it is nesting, it will be difficult to find for it builds beneath the drooping clumps of tussock and there in the shadow its green and blue plumage makes it almost invisible. You may just spot a flash of the scarlet beak, though if the bird is aware of your presence, it will tuck that give-away sign beneath its wing. That is a reaction that it may have developed as a defence not against such late-comers to the islands as human beings but against hawks in search of eggs or chicks. The biggest threat to its survival today, however, comes from a mammal that was introduced from Europe, the red deer. This too

eats tussock, but instead of delicately plucking out individual stems, it devours it in great tufts. This is so destructive that it often kills the plants and at best leaves very little for the takahe. Whether tussock has always been the takahe's mainstay is debatable. It could be that hunting by men during previous centuries drove all the birds out of the more fertile lowland regions where there was better grazing and that the only refuge survivors could find was in these remote and relatively inhospitable mountain valleys.

The most extraordinary of all New Zealand's flightless birds also found sanctuary high in the mountains. It is a parrot, the kakapo. Like the takahe and the moas, it feeds on leaves and like them it has become a giant of its kind, so accommodating the bulky digestive apparatus that is needed to deal with such a diet. It is the largest of all parrots. In order to avoid predators such as harriers or eagles, it has become nocturnal and shelters during the day in burrows. During the night it patrols its territory, nibbling fresh shoots as it goes, and so creates well-worn tracks through the vegetation. It has become New Zealand's native equivalent of a rabbit.

New Zealand provides convincing evidence that birds will return to the ground to find their food and will make their permanent homes there whenever and wherever it is safe for them to do so. In the end, however, the birds' abandonment of flight, after so many years spent in evolving it, has not been successful. Even those that did so seventy million years ago, such as Diatryma and the terror birds, eventually had to surrender their dominance to the mammals and when mammalian predators such as cats, rats and stoats eventually reached New Zealand, the flightless birds there had little defence. Before that invasion New Zealand had at least eighteen species of native birds that were almost if not entirely flightless. Today eleven of those species are extinct and all the remainder are reduced to such tiny populations that they are also in danger of becoming so. The destiny of the feathered reptiles that appeared one hundred and fifty million years ago lay in the sky. There they became, and there they remain today, the undisputed masters.

2

THE MASTERY OF FLIGHT

In the night, on a small offshore island in Japan, birds wait in line to clamber up a tree trunk. They are streaked shearwaters, sea-birds the size of pigeons. They nest in holes in the soft ground and if you are not careful, as you pick your way down through the steeply sloping wood, you either step in one of their holes and trip, or you stumble over a newly emerged bird, in which case it will flounder off in alarm and crash through the blackness into the undergrowth. Like all shearwaters, their legs are placed far back on their bodies. That is the most effective position for propelling them through water. It does, however, mean that they cannot stand up-right. The only way they can get about on land is to shuffle forward in an ungainly fashion with their breasts close to the ground. They make their way down the slope towards a chestnut tree that is particularly suited to their purpose. There they wait their turn to scramble up on to its sloping trunk which is as thick as a telegraph pole. Nose to tail, they inch up it, pushing with their legs, scrabbling with the el-bows of their closed wings. The tree has clearly been used in this way for a long time, for its bark is heavily scratched, down to its red underbark.

The trunk is hollow. Unluckily for the shearwaters, about ten feet up there is a hole in it. Occasionally, one of the climbers blunders into the hole and drops down the interior. Undeterred by this mishap, it emerges at the bottom and rejoins the queue to start all over again. But most of them manage to avoid this set-back and continue to inch higher and higher up the trunk until, some twenty feet above the ground, it bends down towards the horizontal. Having reached that point the birds stop and look about them. They are above the canopy of the surrounding trees and can see out through a gap in the branches towards the moonlit sea. One by one, they open their long narrow wings, lean forward into space and launch themselves into the air. Instantly they are transformed from clumsy halting climbers into

superbly competent aeronauts and away they sail into the black night, off to the sea to fish.

It is the shape of their wings that sustains them aloft. In section, each is thick and rounded at the front edge and tapers, curving slightly downwards, towards the back edge where it has only the thickness of a feather. This aerofoil shape, somewhat like a comma lying on its side, is the almost magical device that keeps a bird in the air. As it glides forward, the air flowing over the upper surface of the wing is deflected upwards so reducing the air pressure just above. The air passing beneath the wing is somewhat impeded by the wing's downward curve, so there the air pressure is increased. With less pressure above and greater pressure below, the wing tends to rise upwards. As a shearwater launches itself off the branch, gravity pulls it downwards and gives it the speed necessary for its wings to function as aerofoils and so keep it in the air. Its laborious climb up the tree has brought its reward.

Most members of the shearwater family nest on cliffs where they can easily shuffle to the edge and fall into space. Usually, there is an on-shore wind which also helps their flight. The Japanese species is unusual in having developed a talent for tree-climbing and this has enabled it to extend its nesting sites farther away from the sea than most other species have been able to do. The wandering albatross, the largest of all sea birds, breeds on oceanic islands where there are seldom either cliffs or trees, so it has to create an air current over its wings in some other way. It uses

*Left: streaked shearwaters
climbing their take-off tree, Japan*

*Above: a waved albatross begins
its take-off run, Galapagos*

the same method as mankind's aeroplanes. Its colonies, like air-ports, have long wide runways that are oriented along the path of the prevailing wind and run right through the colony with nests clustered closely together on either side. To take off, an albatross runs down this clear path into the wind, smacking the ground with its large webbed feet. As it runs, it beats its wings. Each forward-reaching downstroke increases the speed of the air passing over its wings and therefore the strength of the lift they generate, until at last it lifts off the ground. But it can only achieve this if there is at least some head-wind blowing over its wings. On the rare occasions when there is a flat calm on their islands, the albatross are grounded.

One of the heaviest of all flying birds is the swan. A full-grown one may weigh close to 35 pounds. To get into the air a swan requires the most level and smoothest runway in the whole of nature – the still surface of a lake. Even so, taking off demands an explosion of activity. Like the albatross, the swan runs as fast as it can across the surface on its short legs and beats its wings frantically. Spray flies everywhere. As the wings begin to generate lift, so it rises higher above the water. Still it runs, its webbed feet leaving distinct and separate puddles on the surface until eventually they lose contact with the water. The bird tucks them beneath its tail, like an aeroplane retracting its undercarriage, and at last it rises into the air.

Above: a mute swan gathering speed
for take-off, Europe
– 42 –
Right: a wood duck takes off,
North America

The lift created by an aerofoil is directly proportionate to the speed of the air that passes over it, so the speed that a bird can run on take-off governs the size that it can attain. The swan, taking off in still air, must sprint across the water at a speed of 15 yards a second. Watching one do so, it is difficult to imagine how any large bird could manage to run faster. So it may be that any bird heavier than a swan simply would not be able to take off.

Heavy a swan may be, but it is astonishingly light compared with a mammal. A similar-sized mammal, say a bull-dog, weighs about four times as much. Birds owe this lightness to the many adaptations they have acquired since Archaeopteryx first took to the air 150 million years ago – hollow bones supported by internal struts; a fan of stout-quilled feathers instead of a bony tail; and a horny beak instead of jaw-bones laden with a battery of teeth. But they also have one other feature that is not apparent from skeletal remains. A substantial proportion of their bodies is occupied by air. This is held in sacs. Most birds have nine of them. They lie in the neck, the fore-part of the chest and towards the back of the abdomen. Some even extend into the bones of the wings and legs. These are not only weight-saving devices. Flying demands so much energy that a bird needs a very large supply of oxygen. The air sacs are an essential part of its breathing system and enable it to extract far more oxygen from each breath than a similar-sized mammal is able to get.

A mammal's lungs are, in effect, bags. Air taken in with the breath has to return along the same passage, the wind-pipe, as it entered. And the lung does not

completely empty with each breath. The net result is that only about 20 per cent of the oxygen in a breath of air is absorbed by the mammalian lung. A bird's breathing is much more efficient. When it inhales, the air passes first into its lungs. They are relatively small and lie beneath its spine, moulded against its ribs. It then continues through a number of small tubes into those sacs that lie towards the rear of the body. When the bird breathes out, the air in the rear sacs moves back along another set of tubes to a different part of the lungs. With the next breath, this inhaled air moves on yet again, out of the lungs and into another group of air sacs towards the front of the body. Then, with the next exhalation, that air leaves the bird's body through its nostrils and goes back into the atmosphere. In this way, the air-flow in all of the many passage-ways and sacs of the bird's respiratory system is always in the same direction and the absorption of oxygen from each intake of breath is ultimately almost total.

All flying birds share these weight-saving features. Species smaller than a swan and with stronger legs than a shearwater are light enough to get into the air from a standing start. A dove begins by bending its legs and, at the same time, opening its wings and lifting them above its back. As it straightens its legs and starts its jump, it brings down its open wings with such force against the resisting air that the bird is lifted sufficiently far upwards for the tips of its wings to clear the ground. Now it must raise its wings for the second beat. It twists them at the wrist so that they partly fold, thus reducing their surface area, and the long wing feathers separate, allowing air to stream through them. It continues raising its wings until they are above its back. They are now fully open once again and they clap together. This disperses the air between them, lowering its pressure so sucking the bird upwards and reinforcing the upward push created by the wings as they begin their third down-stroke. By the time the dove completes that, it is well above the ground.

Now the great versatility of a bird's wing becomes apparent. Thanks to the manner in which the feathers slide over one another, the surface of the wing remains perfectly smooth whether the wing is closed or opened or at any position in between. The dove, as it beats its wings, not only pushes itself upwards through the air, counteracting the downward pull of gravity, but at the same time reaches forward with a rowing motion so that it maintains its height and at the same time advances through the air. At no stage during the whole action is the smooth flow of air across the wing surface interrupted by any irregularity that might cause it to break up into turbulent eddies. Were that to happen, the lift created by the aerofoil would be greatly diminished or even lost altogether.

The feathers on a bird's body are also crucial in minimising turbulence. They contour it so that the curve from head to neck to shoulder, back and tail is so gentle that the air flows smoothly over it. The importance of this streamlining is easily demonstrated. Watch an osprey fishing. As it flies above a lake, it is the epitome of

Right: a fantail pigeon gains height

grace and aerodynamic efficiency, beating its wings in an almost leisurely fashion. It spots a fish and dives down to grab it. As it rises with its catch in its talons, it greatly increases the rate of its wing beats. It has to make this extra effort not only because of the weight of the fish but because of the drag the fish creates as the bird carries it through the air. To keep this to a minimum, the osprey adjusts its grip so that the fish's head points forward and its streamlined shape which created the least resistance in water, continues to do exactly that now that it is in the air. But the osprey can no longer draw its feet upwards and tuck them away close to its body. These two handicaps cause so much turbulence and drag that the osprey's flight becomes heavy and laboured and it has to beat its wings very much faster simply to stay in the air.

Boobies and gannets, which also fish, have a way of avoiding this problem. They store their catch within their bodies in a crop, a bag that branches internally from their throat. That way even a large catch creates no more than a smooth swelling which, in terms of drag, hinders them hardly at all.

Beating wings demands such energy that it is clearly valuable to do it in as economical a way as possible. One simple method of achieving that is to stop every now and then. A woodpecker in flight, regularly interrupts its rapid wing beats for a few moments by holding its wings closed tightly against its body. Its forward momentum is such that, without the drag created by its open wings, it continues to

Above: an osprey carrying its prey with minimal drag, Europe

Right: white pelicans flying in formation, North America

shoot forward through the air. But it cannot do this for long. Deprived of the lift created by its wing beats, it loses height and after a few seconds it has to resume flapping. This gives it an undulating, bounding flight.

Only a small bird can use this energy-saving trick. If a bigger heavier bird tried to do so, it would drop like a stone. Nonetheless, even a big bird can economise on its wing beats. If it stops flapping with its wings not closed but open, their surface area is big enough to impede its fall and it will glide. Pelicans regularly do this. How long they can glide depends on how high above the ground they are, how much height they can afford to lose, and how fast they are travelling.

White pelicans have a special additional way of saving energy. They take advantage of the turbulence in the air created by their companions. The high pressure air created below a wing by its aerofoil shape leaks round the tip of the wing into the low pressure area on the upper surface. This slight upward current in the air remains briefly in its wake. A pelican flying in a group takes advantage of this by flying behind the wing-tip of the bird ahead rather than directly behind its tail. The

wing-tip station also gives it a better view of what lies ahead. So groups of pelicans often take up a V-formation. Furthermore, because the effect of the wing-tip turbulence is at its greatest immediately after the down-stroke of the wing and rapidly fades, the pelicans not only fly in formation, but beat their wings with the near perfect unison of a well-drilled corps de ballet. The only bird of the group that does not benefit from this order of flight is the leader of the squadron and after having done its share, it will fall behind and allow another to take on the job. Geese also fly in this way to form those huge and unforgettable skeins across the sky.

All birds eventually must come down from the skies, if only to lay their eggs and rear their young. To do so, they must first lose speed. That is done most simply by lowering the tail and the rear edge of the wings. Water-birds with webbed feet stretch them out to serve as additional brakes. The swan is so large that it cannot fly slowly without stalling and falling out of control. It hardly ever attempts a dry landing but comes down on an unimpeded stretch of water. Even so, when its webbed feet, thrust forward, hit the surface it is still moving at such speed that it almost disappears in a cloud of spray. As the waves subside, the swan folds its wings, shuffles them a little to get them neatly into position and then swims away with proper swan-like dignity.

Albatross, coming down to one of their great island colonies, land on the same runways they used for take-off. As they approach the ground, they lower their feet as air-brakes and start paddling in anticipation. Only too often, their speed through the air is faster than they can run along the ground. The result, inevitably, is a minor crash-landing as the bird tips forward on to its chest, but it quickly recovers and staggers away to its nest unharmed.

Ducks and geese have a special technique that allows them to drop sharply through the air and land on quite small ponds. It is called whiffling. They tip from side to side with the long wing feathers widely spread so that they separate and air spills through them making a tearing noise. They may even twist on their backs, so that the aerofoil effect of their wings is totally nullified but their momentum is increased. Only when they are a few metres above their chosen patch of water do they straighten out so that their wings can serve as air brakes and slow them sufficiently for a safe landing.

Perching birds have to alight with much greater accuracy than those that come down on flat ground or water. This, perhaps, puts a curb on their size. Certainly none are quite as big. The harpy of tropical America, one of the biggest of eagles, is only half as heavy as a swan. Nonetheless it is a very big bird indeed and it is very difficult for it to slow down without stalling. Yet it has to reduce its speed to zero at the exact moment that it arrives at its perch. This requires the most accurate

control. If the bird decelerates too rapidly, it will drop beneath its perch. If it does not do so sufficiently, it will overbalance as it reaches the branch and topple forward. As the eagle comes in, it fans its tail and pulls it down at an angle to the body to reduce speed and control direction. It lowers the trailing edges of its wings to act as additional brakes. As it loses speed, there is a danger that turbulence will develop on the upper surface of the wings and cause it to stall. It prevents that by raising its alulas. These are tufts of three or four feathers developed by many flying birds on the leading edge of the wings that are attached to the stunted relics of the thumb bones. They let in a stream of air across the wing surface and so maintain a smooth flow over the aerofoil. The bird has now lost nearly all its air speed. It reaches forward with its huge talons, grabs the branch ahead and finally brings itself to a full stop.

Once landed, a bird must service the feathers on which its life in the air depends. Almost all, if they have the chance, will take a daily bath to rid their feathers of dirt, ruffling up their feathers, ducking their heads and beating the water into a drenching spray. Thoroughly soused, they then comb themselves. Their long wing feathers on which they rely for flight are given particular attention. Each may be carefully passed through the beak so that is is cleaned and any separated filaments zipped back together. The sword-billed hummingbird has a unique problem. It alone among birds has a beak that is longer than its body. As a consequence, it

cannot be used as a comb and the bird has no alternative but to preen itself with one foot while standing precariously on the other. Happily, the processes of evolution which led to the development of its extraordinary beak, thus enabling it to reach nectar in the depths of deep trumpet-shaped flowers, also responded to its need to reach its head-feathers by equipping it with disproportionately long legs.

If water is not available, some birds, such as guans, larks, wrens and sparrows will use dust. That, if followed by a vigorous shake, helps to get rid of parasites of which they may have plenty – chewing lice that nibble their feathers, louse-flies, bugs, mites, fleas and ticks that seek to suck their blood.

Herons and parrots produce a powder for their toiletry. It comes from the fraying ends of specialised feathers. In some species, such as pigeons and parrots, these are scattered throughout the plumage. In others, notably herons, they are clumped together in small patches. Exactly what function this powder plays is not fully understood, but it probably helps with waterproofing. Egrets, pelicans and other water birds anoint themselves with oil that they squeeze from a gland in the skin at the base of the tail. Washing, dusting and powdering completed, the feathers are finally put back into their proper positions.

Left above: a violet-eared hummingbird bathing

– 53 –

Above: a white pelican taking oil from its preen gland, Kenya

Left below: a pheasant taking a dust bath

Even with the best care, however, feathers wear out and all birds have to replace them. For most species, moulting takes place over a longish period. Chaffinches take ten or eleven weeks to do so. A few flight feathers are shed and replaced, followed by others so that at no time is the bird unable to fly. But some birds that can find a safe refuge, as ducks and sea-birds can do on water, change all their feathers quickly over three or four weeks, during which time they are completely flightless.

Different habitats and different ways of collecting food demand different styles of flight and specialist aeronautical equipment. The wandering albatross spends most of its life above the open ocean. There the wind blows almost continuously. Simply facing into the wind with its wings outstretched is enough to keep the albatross in the air. Since the lift created by any wing is at its weakest at the very tip, where the high pressure beneath escapes on the upper surface, it is better, aerodynamically, to keep the tip as far away as possible from its body. That can be done by making the wing very long and the wandering albatross has the greatest wingspan of any bird – twelve feet. If the wind is reasonably strong, these huge wings will generate enough lift to enable the bird to travel at an angle to it. Furthermore, wind whips the surface of the sea into waves. As it blows across them, it is deflected upwards. The albatross is able to exploit these upward currents and tack, zigzagging across the face of the wind, from one wave crest to the next. So skilled is it that it can sail for hours on end without a single wing beat. Holding wings outstretched with tensed muscles would, in itself, require energy for most birds. But

Left: a snowy egret preening, Florida   – 55 –   *Above: a wandering albatross gliding,*
Indian Ocean

the albatross does not have to use its muscles. It has a catch-like mechanism in its wing bones that sets its wings in the open position. So it flies continuously for days, weeks, even months with the minimum expenditure of energy. It has no need to drink for it extracts all the water it needs from its food, the dead squid and fish which it finds floating on the surface of the sea.

The carrion to be found on the surface of the land is collected by other extraordinarily accomplished aeronauts, vultures. They do not have the advantage of an almost ever-present wind. Instead, they exploit rising currents of warm air. Such currents, called thermals, are generated by the way in which different surfaces of the land react to the sun's heat. A tract of lush grass absorbs a great deal of heat. A patch of bare rock, however, reflects heat into the surrounding air. So a column of warm air rises from rock patches and continues upwards, high into the sky. A vulture, once it manages to reach such a thermal, circles within it and spirals upwards until, perhaps a thousand feet above the ground, the thermal has lost so much of its heat into the surrounding air that it no longer has any strength.

Once the vulture has gained sufficient height, it can circle for hours, scanning the plains below for a carcass that might provide it with food. Its wings are not long and slim like those of a high speed glider such as an albatross, which rarely has to dodge obstacles above the open ocean or has the need to make accurate landings. Instead they are very broad with a large surface area that enables the bird to take full advantage of air rising from beneath, while being sufficiently short in span to

allow them to dodge trees and bushes every time they have to make a precision landing.

Flying at high speed requires yet another wing shape. The fastest of all birds – indeed the fastest creature in the air, apart from a human being in a machine – is the peregrine. When that dives on its prey, which is nearly always another bird, it first increases its velocity by beating its wings and then, in the last stage of its stoop, draws them back so that it assumes a silhouette like that of a supersonic jet aircraft and reaches a speed over 200 miles an hour.

The kestrel, a close relative of the peregrine, has a very different hunting tactic. It hangs in the air, apparently stationary, while it searches the ground beneath. In fact, the bird is not motionless in relation to the air around it. It is facing into the wind so that it gets enough lift to remain airborne. It spreads its broad tail to supplement the air-catching effect of its spread wings. It also raises its alulas which further reduce the danger of stalling because of turbulence. It separates the feathers at the broad ends of its wings so that little upward jets of air are generated which dispel any turbulent eddies on its upper surface. Carefully adjusting these various controls, it manages to match exactly its forward motion through the air with the speed of the wind and hangs directly above the patch of ground it is scanning for prey.

Only one family of birds can truly hover in still air for any length of time. They are the hummingbirds. They need to do so in order to hang in front of a flower while they perform the delicate task of inserting their slim sharp bills into its depths to drink nectar. Their thin wings are not contoured into the shape of aerofoils and do not generate lift in this way. Their flying technique differs from that of most other birds' as radically as helicopters do from fixed-wing aircraft.

The long bones of their wings have been greatly reduced in length, and the joints at the wrist and the elbow have lost nearly all their mobility. Their paddle-shaped wings are, in effect, hands that swivel at the shoulder. They beat them in such a way that the tip of each wing follows the line of a figure-of-eight, lying on its side. The wing moves forward and downwards into the front loop of the eight, creating lift. As it begins to come up and goes back, the wing twists through 180 degrees so that once again it creates a downward thrust. If the lift produced on each loop of the figure-of-eight is equally strong, the bird will remain stationary in the air. A slight alteration in the twist, changing the angle at which the wing moves downward, is enough to move the bird forwards – or even backwards.

This manner of flying demands a great deal of energy. Nectar, the humming-bird's food, is the biological equivalent of high-octane fuel but even so a humming-bird consumes such quantities that it may need to refuel as many as two thousand times a day. Even at rest its body needs a great deal of fuel simply to keep ticking over. Part of this is used to keep its flying muscles at a high temperature and ready for instant take-off. But when night comes and it is unable to see to fly, those muscles are not used and that heat is not needed. So in the evening when a humming-bird arrives on its roost, it deliberately ruffles its feathers and allows its body to cool. During the day, its heart beats between five hundred and twelve hundred times a second. Now it slows down so much that its throb is virtually undetectable. Nor does the bird appear to be breathing. In effect, it is doing what a hedgehog does when winter approaches. It is hibernating. A hummingbird, however, has to hibernate three hundred and sixty five times every year.

The hummingbird's method of flying does have a major limitation. The smaller a wing, the faster it has to beat in order to produce sufficient downward thrust. An average sized hummer beats them 25 times a second. The bee hummingbird from Cuba which, being only two inches long, is indeed scarcely bigger than a bumble bee, has to do so at an astonishing two hundred times a second. There is a limit however to the speed at which an electric signal can pass down a nerve in order to trigger a muscle. The bee hummingbird's wing-beat is on the edge of that limit. There can be no smaller flying bird.

There are, of course, smaller flying insects. They, however, have had to adopt a fundamentally different mechanism for beating their wings at high speed. Their bodies are encased in a horny external skeleton to which the wings are fixed. They

Left above: a broad-billed hummingbird about to feed from blossom, Arizona
Left below: a white-bellied emerald hummingbird in nocturnal torpor, Belize

are able to make this vibrate with a relatively simple contraction of a muscle in the same sort of way that a sharp blow will make a tuning fork vibrate. They therefore have no need to send separate nerve pulses to initiate every beat.

The power of flight, demanding though it is, has made birds the swiftest moving creatures on the planet. The fastest animal on earth is a cheetah which recent research suggests cannot exceed 50 miles an hour even in short sprints. The fastest fish in the sea, the sailfish can, exceptionally and over a short distance, achieve a speed of 65 miles an hour. But the spine-tailed swift in level flight has been credited with a speed of 100 miles an hour. Flight has also enabled birds to overcome all the physical obstacles that restrict the movements of land-living animals. They, to a degree unequalled by any other kind of animal, are able to travel anywhere to find the conditions that best suit them at any particular time, avoiding seasonal bad weather and visiting places where food is suddenly but only briefly available.

The high Arctic is just such a place. For at least six months of the year, conditions are crushingly hostile to life. For much of that time, the sun does not rise above the horizon. Even when it does, its rays are so low and glancing that they give little heat to the snow-covered ground. Without light, plants cannot grow; and without plants, herbivorous animals cannot feed and predatory animals have no prey. Nonetheless, a few animals and plants manage to live here permanently. They do so by reducing their life processes to the bare minimum during the dark winter months and concentrating all their activity into the short summer. As the sun begins to clear the horizon in spring, temperatures begin to rise. The snow melts and reveals stunted heather, willows, saxifrages and poppies. The ice on the bog pools disappears and cotton grass begins to flower. Lemmings that have been hibernating deep in burrows beneath the snow, venture out and start ravenously to crop the leaves and make good the deficits of winter. And from the pools, insects haul themselves from over-wintering pupae and swarm in such numbers that the air hums with their droning. By high summer, the Arctic animals are more continuously active than any below the Arctic circle, for now it is light twenty four hours a day.

There is food and to spare for birds. Poverty has become plenty and birds fly up from the south to benefit from the glut. Among them are snow geese. They establish colonies of a hundred thousand pairs or more among the clumps of coarse tussock grass. Had they done that farther south, they would have attracted stoats and weasels, foxes, wolverines and wolves and other predators. But such four-footed hunters do not appear because they cannot make the long journey northwards as swiftly and easily as birds. The only mammalian predators that birds have to fear up here is a small permanent population of Arctic foxes. With so many geese nesting, these foxes are faced with far more eggs than they can possibly consume. Those they cannot eat immediately they bury to be devoured later, but even so, the foxes are so few that they have little effect on the huge assembly of snow geese.

By late summer most geese have four or five young which are already fledging. Unlike their parents, which are pure white except for the black tips to their wings, the young have grey upper parts. All over the tundra, these goose families are busy feeding, the adults digging for roots of bull-rushes and the young nibbling the tender tips of the marsh grass leaves. They all put on weight fast. They need to. They cannot stay much longer in the face of the approaching winter. They have a long journey ahead of them.

Two thousand miles farther south, in the Canadian woodlands, the ruby-throated hummingbird, almost as different in size and flying skills from the snow goose as it is possible for a bird to be, is behaving in exactly the same way. It travelled up from the south to take advantage of the summer flush of flowers. The supply of nectar was so lavish that it was able to feed not only itself, but its brood of young. But now that the plants have been pollinated, their flowers are withering and falling. So the hummingbirds and their young must also embark on the long journey south.

Above: a snow goose parent with its fledgling young, Canada

Raptors – hawks, buzzards and eagles – also came to these woods, hunting for voles and other rodents as well as preying on smaller birds and their newly hatched young. But the voles are beginning to disappear below ground to hibernate and most of the small birds are starting to depart southwards. The raptors must follow. Six hundred and fifty species of birds feasted and nested in the lands of North America during the summer. Five hundred and twenty of them, come the autumn, prepare to retreat in the face of the increasing cold.

Different birds, with different flying abilities, have developed different travel strategies. The snow geese, since they are large comparatively heavy birds, have to fly fast to keep airborne. That, in itself requires them to carry substantial food supplies as fuel on even short flights. Even at the best of times, they have little reserves of power to spare. They will have to make frequent stops on the way to refuel, sometimes spending several days feeding intensively before they have taken on enough to resume their journey.

The raptors are more fortunate. Gliding uses only a twentieth of the energy required for beating wings. They will rely on thermals to take them to altitude and then they cover as much distance as they can in long shallow glides that are almost

effortless. So they will wait for a good warm day before they set off. They also know particular places where the thermals are powerful and reliable. Hawk Mountain in Pennsylvania is just such a place and in September, just before the woods begin to flush red with autumn colours, thousands of raptors begin to assemble.

Waders look small and frail, but they are among the most competent and accomplished of travellers. They do not have to rely on warm weather as hawks do and, unlike geese, they can carry significant reserves of fuel. They accumulate it in the form of fat and feed so voraciously on the mud flats of the coast that in a few weeks they almost double their summer weight. These reserves are even bigger than that statistic might suggest, for many of their internal organs, including their brain and their guts, shrink in size to accommodate this additional fuel and save weight.

So all over North America, the mass migration begins. Flying during the day under a blazing sun can bring a risk of serious over-heating for birds that beat their wings non-stop. Geese are able to avoid the danger by travelling at night. The families keep together by calling to one another as they go. Smaller birds – thrushes, flycatchers and many others – are also night flyers. Raptors, however, have no option. They must travel by day when the thermals are strong but the problem for them is

– 63 –

Above: skeins of migrating snow geese, North America
Following pages: turnstones, knots and sanderlings assembling before their migration, North America

not so severe since so much of their time in the air is spent gliding. The ruby-throated hummingbird also flies during the day, for it is only then that its fuelling stations, the flowers, are open. It follows traditional routes with extraordinary precision. The same ringed bird will appear from the north in the autumn, year after year, to drink from the same flowering bush.

Waders – sandpipers, sanderlings and knots – find much of their food on the sea-shore and mudflats, so those that leave the Arctic tundra and travel down the east side of the continent, stick to the coast of Hudson's Bay as long as they can. They then make a long overland trip across southern Canada down to the coast of New England. There they pause, restock their fat reserves and, in many cases moult so that they have a new set of feathers with which to tackle the ocean crossing to South America. Although they spend their lives close to water, they are not able to settle on its surface or to swim. Their two thousand mile sea passage will have to be non-stop.

Those birds that have been travelling overland down the centre of the continent, many following the broad path of the Mississippi valley, eventually reach the Gulf of Mexico. The shortest crossing, from Texas to the Yucatan peninsula, is a journey of five hundred miles. To go round the west coast, by way of Texas and northern Mexico, is three times farther. To take the eastern route, down Florida and

Above: a flock of European knots

onwards by way of Cuba, is also a much longer journey and involves several major sea crossings from one Caribbean island to another.

The raptors once more have no alternative. There are no thermals over the sea so they have to take the long western route, overland. Ducks, plovers, nightjars and swallows opt for the eastern route, along the islands of the Caribbean. The tiny ruby-throated hummingbird, almost unbelievably, tackles the sea crossing directly. Its cruising speed is about 27 miles an hour, so if conditions are favourable, it can make the transit, non-stop, in around 18 hours. But the passage is a formidable one and it taxes the hummingbird to the very limit of its endurance. A head wind, even a mild one, may hamper it so severely that it will never reach the far shore and perish at sea.

How do birds find their way on these immense journeys? There is no single answer. Each species almost certainly uses several techniques. Some follow major geographical features – the Appalachian Mountains, the Mississippi valley, the coastline of Hudson's Bay. Night flyers navigate by the stars, perhaps by recognising the point of the night sky around which the constellations revolve which is conveniently marked by the pole star. As a consequence, on cloudy nights they may go astray and sometimes get quite lost. Day-flyers use the sun, though this is more difficult since the sun moves so swiftly. If they do use it, they must also have an internal clock to give them an accurate idea of the time. More mysteriously, many if not all birds are able to sense the magnetic field of the earth. That can be demonstrated by fitting a group of them with slender rods of iron. Some are given rods magnetised in a way that obscures the earth's magnetism. Birds with these get lost. Others are given unmagnetised bars, and these find their way perfectly well. Microscopic grains of an iron oxide, magnetite, have been discovered in the brains of some birds. It may be that they are part of the mechanism of this sensing system. Some birds, like the geese which travel in families, undoubtedly learn their traditional routes from their parents and pass on that information by example to their own young. Young European cuckoos, however, are always abandoned by their parents, yet they are able to find their way down to the savannahs of Africa totally unaided. They must have inherited their route-finding skill genetically.

By October, the great journeying is coming to an end. The snow geese have not tackled the sea-passage across the Mexican Gulf. They have settled for the winter around the Mississippi delta and farther to the west in the northern part of Mexico. The ruby-throated hummingbird has reached southern Mexico and Panama. Hawks, warblers and nightjars, ducks, plovers and terns are all settled in their warm South American quarters for the winter. Bobolinks, relatives of the oriole, have made the longest journey of all North American birds. They bred in northern Canada and not only crossed the Caribbean but continued south and after a journey of five thousand miles, reached their final destination on the pampas of

Argentina. All these migrants will remain in the south while their summer homes in the north are covered by snow and ice. Then, after a few months, their internal clocks and the changing season will tell them that the time has arrived for them to return once more to feast in the lands of the north.

The pattern of migration is particularly clear in the Americas, but the phenomenon occurs all over the world. European swallows having nested in England fly across Europe, over the Mediterranean either by way of the Straits of Gibraltar or down the long peninsula of Italy and on to the island of Sicily, then across the ferociously hot sands of the Sahara and down to the grasslands of South Africa. In Asia, bar-headed geese that nested on the high plateau of Tibet fly right across the Himalayas, sometimes travelling as high as 2,500 feet, to winter in India and knots from the high Arctic fly south along the coasts of Japan and Vietnam and may even cross the South China Sea to winter on the southern and western coasts of Australia. In the southern hemisphere, the autumn journeys, of course, are in the opposite direction. The biggest of all humming birds, the Patagonian giant, flies up from the bleak mountains of Chile to the lush forests of Ecuador and long-tailed cuckoos travel from New Zealand across the Tasman Sea to tropical Australia.

Above: Arctic terns

The rewards for these arduous journeys are clear, even if they are hard-won. By making them, the birds are able to reach resources of food in parts of the world where they could not live permanently. How they came to know of the existence of these resources so far away and across such immense obstacles is less clear. It lies in the past.

The earth sometimes undergoes periods of cooling. There have been two Ice Ages within the last 150,000 years. During those times, the areas where birds could live were limited to a band of lands on either side of the equator. But as the earth warmed again, those regions began to expand to the north and to the south. As they did so, the birds followed. Each year, the journey to reach the summer feeding grounds became longer, but only marginally so and the birds were able to keep pace with the changes, learning the way and developing the techniques of navigation. If the world continues to warm in coming centuries, as it seems possible that it may, then snow geese and bobolinks may eventually make even longer journeys than they do today.

The power of flight has enabled birds to exploit every part of the planet in a way that cannot be paralleled by any other group of animals. No physical obstacle has totally defeated them. They have crossed the highest mountain range. They have traversed the widest ocean. The longest of all annual journeys is made by Arctic terns. Every August, they leave their summer quarters in the north and start to fly south. Those in Arctic Canada and Greenland cross the Atlantic and meet others that have come from Arctic Russia around the coasts of western Europe. They travel onwards skirting the coast of West Africa. Some do so all the way to the Cape of Good Hope. Others cross the Atlantic again to the eastern coast of South America. A third group travels down the eastern margin of the Pacific from the north west coast of Canada, down across the seas off California, Peru and Chile to Cape Horn. All three groups then cross the Great Southern Ocean to Antarctica. At the height of the northern summer they experienced daylight for twenty four hours of the day. Now they do the same in the Antarctic so they see more of the sun in a year than any other animal. The annual round trip could be as much as 25,000 miles.

Terns can come down briefly on the sea or perch on icebergs, so they are able to get some rest. The European swift, however, does not even do that. Its legs are of even less use to it than the those of the shearwater, for they are reduced to little more than clusters of four curved needle-thin claws. If a swift comes down to the ground, it finds it almost impossible to take off again. It nests in holes in cliffs and, predominantly these days, in the attics and lofts of houses which it manages to enter through grills and other spaces below the eaves. To make its rudimentary cup-like nest, it snatches feathers and grasses that drift in the air and cements them together with its sticky spittle. It drinks on the wing, dipping down to make a shallow

dive through the surface of a pond. It feeds in mid-air entirely on flying insects. It sleeps in mid-air too, rising in the evening to heights of 6,500 feet and drifting in the wind with only occasional flickers of its outstretched wings. It even mates in mid-air. The male clings on the female's back and united the two descend for a few seconds in a shallow glide. When a swift, young or adult, leaves its nest in early August, bound for Africa, it may not touch down again until it returns to its nest site nine months later.

An individual swift is known to have lived for as long as eighteen years. In its lifetime, it must have flown some four million miles. That is the equivalent of flying to the moon and back eight times. The swift, truly, is the most aerial of animals.

Above: a swift on its nest, Europe

3

THE INSATIABLE APPETITE

Powered flight requires fuel. For birds, that fuel is food; and since flight demands that bulk and weight should be kept to a minimum, the more compact and powerful that fuel is, the better. Seeds have both those qualities. The nourishment they contain is there to enable a developing seedling to build a stem and leaves so that it can start to manufacture food on its own account, but that same nourishment can also feed birds and it is so rich and so conveniently packaged that many birds eat little else. Since it is of no benefit to a plant to have its seeds destroyed in the stomachs of birds, many plants armour their seeds to prevent that happening. Birds, in response, have evolved special tools and strategies to ensure that they can continue to plunder this valuable food supply that suits their needs so well.

The variety and versatility of tools that birds might use for this purpose is greatly reduced by the modifications made for flight so many millions of years ago. Front legs transformed into wings can no longer grasp and rip. And a horny beak is not nearly as good at chewing and grinding as a pair of bony jaws armed with powerful teeth. All a beak can do, mechanically, is to open and close.

Nevertheless, it is extraordinary how versatile and effective a beak can be. Keratin, the substance from which it is made, seems to be easily moulded by evolutionary pressures. Consider the finches. The chaffinch has a general all-purpose beak. It is approximately conical when closed, has a sharp point and it is neither extremely long nor short. It enables its owner to grasp insects, spiders, caterpillars and berries. And it also serves very well as a pair of tweezers with which to pick up seeds.

But armoured seeds defeat it. Early in the finches' family history, one population must have emerged with slightly heavier bills which were somewhat better at cracking the shells of seeds. Some plants, in response, produced seeds with slightly

thicker armour. But the bigger-billed finches found that it was increasingly worth-while to concentrate on seeds which others in their family, without such heavy-duty implements, could not collect. As they specialised in this way, so they developed increasingly stout bills.

So now the greenfinch has a heavier beak than the chaffinch. It can tackle seeds that would certainly defeat a chaffinch and has no difficulty in cracking sunflower seeds. Cherrystones, however, are stronger still. But the hawfinch has an even more powerful beak with a pair of ridged knobs at the base of each mandible which will grip a cherry-stone. The muscles which close this beak are exceptionally large and extend right round the skull. When they contract, they can exert a force in excess of a hundred pounds and that is quite sufficient to crack a cherrystone in two and allow the hawfinch to steal the seed within.

Other members of the finch family have developed different specialisations. The teasel protects its seeds by covering them with long sharp spines. The goldfinch, one of the smaller finches, has evolved a bill with longer more slender proportions than any of its cousins, so that it can perch on a teasel head and reach between the spines to take the seeds without risk of spiking its head or its eyes. A pine tree holds its seeds between the scales of its cones. As the cone ages, it becomes hard and woody and the seeds fall to the ground where they may be eaten by mice and squirrels as well as many kinds of birds. But the crossbill collects them before that can happen. Its mandibles are crossed at the tips. Holding its head on one side, it inserts this strange implement between the scales of the cone. The upturned point of the lower mandible pushes the scales apart so that the bird can either hook out the

Above left: a chaffinch eating grain, Scotland   – 72 –   *Right: a goldfinch on a teasel, England*
Above right: a hawfinch with
a sunflower seed, France

seeds with its upper mandible or scoop them out with its tongue. It works so swiftly and dexterously that it is not easy to analyse exactly how it achieves its results. It is certainly the case, however, that a crossbill without crossed mandibles cannot extract seeds from a closed cone. But this bill, like many specialised tools, has its limitations. A crossbill can no longer pick up seeds from the ground as other birds do.

Acorns, the seeds of the oak tree, are particularly nutritious. They have a shell that is too tough for any finch to split. They are also very large and too big for any finch to swallow. Even a comparatively big bird like a jay cannot eat an acorn whole. It has to take it to a favourite crevice, in a stone or a branch, wedge it in and then, using its beak like an ice pick, first split away the shell, and then break up the seed itself into fragments. The oak tree's most effective defence for its seeds is not, however their armour, it is their number. In late summer, a mature tree may carry as many as ninety thousand acorns. That is far in excess of what all the acorn-eating birds around at the time can possibly consume. Some, surely, must survive. The jay does its best to make sure that they don't.

Pigs and deer, which also relish acorns, eat prodigiously during the glut and store the surplus nourishment as fat. Birds cannot do that, for they cannot afford to fly around carrying such extra weight. They have to have a different method of storage. The jay buries them individually. It digs a hole, puts in the acorn and then

Above: a white winged crossbill, Canada   – 74 –

carefully covers it up so that other animals are unlikely to notice it. It then takes careful note of the landmarks around each cache so that it stands a chance of finding the hidden seed some time later. Prominent trees, fallen logs, boulders, and fence posts can all act as signposts. The birds may even place small pebbles and stones nearby which perhaps act as markers.

Many species of birds store seeds in this way and their industry as well as their memories have proved to be very remarkable indeed. Willow tits hold the record for the number that an individual bird will store in a day – over a thousand. The nutcracker, a European relative of the crow, collects the greatest number to be gathered in a single season – up to 100,000. It also has the longest demonstrable memory. Researchers have observed one of them retrieving seeds nine months after storing them.

But even the most tenacious memory is not perfect. Jays certainly do not remember every acorn they bury. And that, of course, is a reprieve for the oak, for then its acorns, having been neatly planted in the soil and covered up by the jay, are out of the sight of other seed-robbers and in just the sort of conditions they need for successful germination.

Woodpeckers, although most are primarily insect-eaters, also eat and store acorns. Exceptionally, one North American species, the acorn woodpecker, as its

Above: a jay with an acorn, Sweden

name suggests, eats little else and its storage techniques are so efficient that it provides no service of any kind to oaks. It uses a tall and very often dead tree. Even a telegraph pole will serve. During the summer, it spends a lot of time chiselling small funnel-shaped holes into it. In late summer, the birds bring acorn after acorn to the tree and hammer them into the sockets, one by one. Acorns are not all the same size, and neither are the sockets, so finding exactly the right hole for each one is not easy. If the hole is too small, then the acorn will be damaged if it is forced in and will later rot. If it is too big, then the acorn will be loose and easily stolen. So the initial placing can take time and involve a great deal of trial and error. And the work does not end there. Acorns, as they age, dry out and shrink a little, so they may have to be re-housed as time passes. A really big store may contain as many as fifty thousand. Curating such a collection and guarding it against thieves is far beyond the ability of a single bird or even a single pair and these rich treasuries are tended by a whole family of woodpeckers.

The birds' lack of teeth not only limits the way in which they can gather food. It also minimises the degree to which they can process it before they swallow it. Seed-eaters can remove inedible shells and even break the kernels into large fragments, but little more. Chewing and munching is beyond them. That has to be done elsewhere in their bodies. Their stomachs have two chambers. The front one has abundant glands in its lining the secretions of which process the food chemically. The rear chamber deals with it physically. This is the gizzard. It is shaped rather like a flat circular purse with walls that are thick and muscular and have a grooved and pleated inner surface. The muscles contract rhythmically, squeezing the walls together and grinding them against one another so that whatever lies between them is broken up and thoroughly mixed with the digestive juices secreted by the front chamber. All birds have such a mill, but seed-eaters need one that is particularly powerful to break away the hard shells of the seeds. And to increase its efficiency, they fill it with grit.

Grit in the gizzard is not as heavy as a mouthful of molar teeth, but even if it were, it is better placed, from an aerodynamic point of view, for it lies in the centre of the body and not right at the front end of the head where it would make a flying bird very nose-heavy. Even so, its weight is a significant load. The bearded tit only eats seeds for a short period each year. When spring comes, its diet changes to insects. Such soft food does not need such intensive processing, so in spring the tit gets rid of its grit and its gizzard shrinks.

Birds that do not spend much of their time in the air, such as chickens and turkeys, are less ruled by considerations of weight. They can afford to have particularly large gizzards and continually renew the grit within them. A domestic goose carries around about an ounce of it. Those birds that have totally abandoned flight have even heavier internal mills. The ostrich is sometimes thought stupid because it will pick up and swallow large and plainly inedible objects, especially if they are bright and glittering. In fact, it is prudently renewing the tools of its gizzard. The biggest gizzards of all were probably those of the moas, the giant extinct flightless birds of New Zealand. Clusters of stones, varying in size from grains of sand to pebbles four inches across, have been found lying within the ribs of an excavated moa skeleton. Their rounded shape and high polish make it clear that they were once within a gizzard. One group belonging to a single bird weighed as much as ten pounds.

Tough shells are not the only protection that plants provide for their seeds. Some also load them with poison. One of the most virulent of all natural poisons is strychnine. That comes from the seeds of a South American liana. Other seeds, while not normally lethal, are nonetheless so unpleasant that most birds will leave them well alone. Several trees in the Amazonian forests are like this. Macaws, however, specialise in dealing with seeds, even the most heavily protected and intractable. Their huge hooked bills are capable of cracking the toughest nut and they are not deterred from swallowing the kernels just because they are likely to cause severe indigestion. Having fed, the birds fly off to particular places where the rivers are flanked by cliffs of clay over a hundred feet high. There they gnaw away at the clay and swallow it in large quantities. The clay, like the doses of kaolin we ourselves may take to cure food poisoning, absorbs the toxins in the seeds and so the birds can digest the nutritious part of their meals without acute stomach pains.

Seeds are not the only parts of plants that are stolen and eaten by birds. Some birds specialise in drinking sap, the liquid loaded with sugars and just a dash of amino acids, that plants manufacture in their leaves and circulate through their stems and trunks. Relatives of the woodpeckers called sap-suckers specialise in doing this. They drill pits in the trunks of living trees that penetrate the bark and break into the sap-carrying tubes that lie just beneath. Each pit they make is inclined slightly downwards so that a pool of sap accumulates within it. The birds are then able to lap it up with their tongues which are covered with coarse bristles and are almost brush-like. Insects also come to drink the sap and settle around the pits in large numbers. The birds lap them up too and so add a little protein to their sugary diet.

The tree defends itself against this blood-letting by growing a dry covering over the wound. It is possible that sap-suckers have a kind of anti-coagulant in their

Above: parrots and macaws eating clay on a river bank, Peru

saliva which hinders the tree in sealing up its wounds, but even so, after ten days or so, the flow of sap comes to an end and the sap-suckers have to drill new pits. During the nesting season, when the birds have to feed chicks as well as themselves, each adult may drill as many as four pits a day, so that before long a band of pits, a dozen or so rows deep, stretches round the tree trunk and may completely encircle it. The loss of sap may then be so severe that the tree slowly dies.

A few birds also rob plants of their leaves. Leaves, however, are not really a very suitable food for a bird. They are very bulky and contain far less nourishment, pound for pound, than seeds or sap. Mammals that specialise in eating leaves, such as cows and antelope, horses and rabbits, have particularly large stomachs in which they hold their meals while they are given long and thorough treatment with digestive acids. Rabbits give everything they eat a second processing by the simple expedient of collecting their pellets of dung as they emerge, usually at night, and swallowing them again so that they pass through their stomach and gut for a second time. Horses recruit the aid of bacteria. Cultures of micro-organisms flourish in their gut just behind the main chamber of the stomach and help to ferment the chewed leaves and extract sugars from their cellulose. Cows use a more complex method. Each mouthful of leaves, having been given an initial chew, goes down into a special chamber in front of the main stomach where it is stored. Then later in

Above: a red-breasted sapsucker,
North America

Right: hoatzins, Peru

the day, when the cow is resting after a hard morning's grazing, each lump is brought up again to be given a second chew.

No bird has managed to parallel such techniques. None collect its faeces in mid-flight and none manages to chew its meals twice. Nonetheless, a few birds do manage to live on leaves. Geese do. Their technique is fundamentally different from a rabbit's or a cow's. Instead of spending a long time extracting the maximum nourishment from their meals, they simply take what is most easily and immediately absorbed - the tender growing tips of the grass-blades - pass it swiftly through their digestive tracts and get rid of what remains as soon as possible. The disadvantage of this practice is that to get enough nourishment, they have to eat intensively and non-stop for long periods. They can pick a hundred leaves of grass a minute. They will graze morning and afternoon. If a full moon enables them to keep watch for predators, they will continue eating through the night as well. And they defaecate as fast and almost as continuously as they eat.

One bird tries to extract rather more from leaves than geese do. This is the hoatzin, the odd South American bird whose nestlings have claws on their wings. It lives in swamps among mangroves and dense stands of moka-moka, a kind of giant water-living arum. It spend its mornings ripping off pieces of arum leaves with its beak and swallowing them. They accumulate in its crop, a huge muscular bag opening from its throat just in front of its stomach. When this is stuffed full to

bursting, the bird lumbers into the air and laboriously flaps its way to a roost where it sits with its distended breast resting on the perch between its feet. Now bacteria and other microbes in its crop get to work fermenting the leaves, while the muscular walls of the crop squeeze and heave, reducing the leaves to a fatty smelly porridge. Forty-eight hours later, the remains of the meal are excreted, smelling – somewhat unusually for a bird – a little like cow-dung. Not surprisingly, one of the local names for the hoatzin is 'stink bird'.

Leaves, seeds and sap are, of course, essential parts of a plant's anatomy, and in taking them birds are inflicting real damage. But plants also produce substances for no other purpose than to be eaten by birds. That being the case, they are neither hidden away, nor armour-plated nor filled with poison. They are flaunted and made appealing to the eye and attractive to the taste. They are bribes which, if taken, will induce a bird to provide a service for a plant, either by collecting its pollen and transferring it to another, or by carrying its seeds away undamaged and depositing them some way from their parent.

Birds do not have a particularly discriminating sense of taste. A parrot has only about 350 taste buds in its tongue, whereas a rabbit has around seventeen thousand. Nonetheless, birds clearly prefer some flavours to others and many are fond of sweetness. Plants can therefore attract them with nothing more than a very dilute solution of sugar. This is nectar and they secrete it from small glands usually in the centre of their flowers. The coloured petals advertise its presence and the flowers are so shaped that when a bird comes to collect the nectar, it brushes against the stamens and collects a load of pollen. When it then feeds from another flower, some of that pollen will brush off and the plant's purpose of cross-fertilisation will have been achieved.

Many birds in the temperate lands of the north have a taste for nectar, but no flowers bloom during the winter, so none of the birds that live there can live on nectar alone. Things are different in the tropics. Flowers can be found there throughout the year, so birds can make nectar their mainstay and develop specialised equipment with which to collect it. In Australia lorikeets, members of the parrot family, have developed tongues which are covered with little fleshy hairs with which their owners are able to brush up the nectar. African sunbirds use different instruments – long tongues inside slender, slightly curved beaks with which they can reach into the depths of a flower and suck up the sugary fluid.

Just as it is advantageous for a bird to specialise in taking one particular kind of food, so it also suits a plant to have its own exclusive staff of employees who, when it is in bloom, will only visit other plants of the same species. Pollen delivered to a different kind of plant is pollen wasted. This mutual interest in specialisation has led to very close partnerships. Many flowers have shapes that allow only one kind of bird with one kind of beak to drink from them. The golden-winged sunbird lives

Right: a malachite sunbird gathering nectar from an aloe, South Africa

on Mount Kenya and has a long strongly curved bill that matches the shape and dimensions of the flowers of the wild mint so closely that the bird can swiftly dip it
into the flower and drink. But the plant is not prodigal with its wages. It only provides the sunbird with a tiny sip in each flower. If it were more generous, the sunbird could drink its fill and then take a rest. As it is, the sunbird has to hasten away
to visit other flowers and keep itself topped up with fuel. So the plant succeeds in
getting its pollen spread around widely and speedily. Every day, a golden-winged
sunbird visits 1,600 mint flowers.

Few plants flower continuously, so as the year progresses sunbirds have to
transfer their custom from one species of flower to another and there will always be
a few who will try to take nectar from flowers with which they do not have an established partnership. But they are unlikely to be able to do that effectively or
speedily. The malachite sunbird, which normally feeds on the flowers of aloes, has
a straight bill and can only reach the nectar held by the mint by forcing its beak
down the curved throat of the flower with a series of jabs. Even then, it often fails
to find the entrance of the little chamber that holds the nectar. The golden-winged
will chase such a poacher away from its territory. That suits the mint, for if the
malachite did collect its pollen it would in all probability deliver it to another kind
of plant where it would have no value.

The variable sunbird, which also lives on the mountain and feeds on flowers
which have little depth to them, has even greater difficulty in reaching the mint's
nectar. Its beak and tongue are altogether too short. The only way it can reach the
nectar is to stab a hole in the base of the flower from the outside. This is a total loss
to the mint. Its flowers are damaged and since the bird never puts its head inside the
flower, it never takes any of the mint's pollen anywhere. But fortunately for the
mint, its regular employee, the golden-winged sunbird, is vigilant in guarding it
against such smash and grab raids.

In South America, the relationship between birds and their nectar suppliers has
become even more elaborate and extreme. Hummingbirds, which are the most
abundant and specialised of the world's nectar drinkers, are superficially similar to
the sunbirds of Africa. Both are about the same size; both are brilliantly coloured
with iridescent plumage of great beauty; and both feed by inserting their thin delicate beaks into flowers and collecting the nectar with threadlike tongues. But the
two families are not at all closely related. The resemblance between them is due to
the fact that both follow the same way of life.

South American plants, however, make even greater demands on the capabilities of their partners. African plants that are pollinated by sunbirds tend to bear
their flowers on twigs or stout stems, so birds can perch beside them or even on
them. In some cases they can even by reached by bigger animals such as monkeys
and squirrels that might have a fancy for a meal of petals nicely sweetened by

nectar. South American plants, by contrast, suspend their flowers on long delicate stems so that they can only be easily approached from the air. Hummingbirds, in response, have developed their own unique way of flying that allows them to position themselves in mid-air directly in front of a blossom and insert their bills deep into a flower with the greatest accuracy. It is a skill that no African sunbird can match. And these partnerships between particular species of plant and bird have also reached more extreme levels of intimacy and exclusivity. Some hummingbirds have bills that are sharply curved downwards. One has a beak that is turned up. Some, like the little tufted coquette, have a tiny bill as straight and as sharp as a needle. And the hugely disproportionate 3-inch long beak of the sword-billed hummingbird, longer than its body, is a device that enables it to be the only member of its family that can sip nectar from the deep trumpet flowers of the Datura plant.

Transporting pollen is not really a huge labour. The birds that do so show no sign that they are even aware of the service they are providing and they are certainly not inconvenienced by the pollen's weight which is infinitesimal. But the other service for which plants require birds is a very different matter. They employ them to carry their seeds and the wages for that job have to be very substantial to make it worthwhile.

And they are. The avocado offers a rich oily flesh wrapped around its very substantial seeds. This is the favoured food of that most spectacular of Central

Above: a woodstar hummingbird feeding from hibiscus, the Bahamas

American birds, the quetzal. The bird is about the size of a large pigeon, but has an immensely long and glorious tail. The male is clothed above in glittering iridescent green and below, on his breast and belly, in scarlet. The female is similar, though her colours are not as bright and her tail is not as long. Not surprisingly, this splendid creature was regarded by the human inhabitants of the area, the Maya, as being sacred and it is still today the national bird of Guatemala. When the breeding season comes and great quantities of food are needed for raising a family, a pair of quetzals will deliberately choose a nest hole close to an avocado tree from which they can collect rich meals every day.

An avocado's fruit hangs on a long stem. The quetzal flies at it with its beak agape, seizes the fruit in mid-air and relies on its weight and speed to rip the fruit away from the branch. It then flies off to a favoured perch where it swallows its prize. The fruit is so big that the bird is only just able to get it into its beak and down its throat. Digestive processes in its crop strip off the nourishing rind and the bird is left with the hard indigestible stone. This is far too heavy an object to carry around in flight for long. It is also too large, doubtless, to pass through the bird's body and out at the other end. So in due course, the bird regurgitates it. During the nesting season, it often does that inside its nest hole. Seeds dumped there are wasted, from the avocado's point of view, but at other times they may drop to the ground sufficiently far from their parent that any young plants that spring from them will not compete with it.

Left: a resplendent quetzal, Guatemala — 87 — *Above: a keel-billed toucan feeding on pawpaw, Belize*

Smaller seeds are dealt with in a different way. A toucan tackling a big fleshy fruit like a pawpaw, a redwing gorging on holly berries, a waxwing collecting rose hips from a hedge, all swallow the fruit either whole or in great lumps and then, after the relatively small seeds have passed unharmed right through their digestive tract, they excrete them with the rest of their droppings. You might think that this lengthy process would be more beneficial for the plant since it allows time for the bird to fly farther away from the place where it has made its meal. But here the parsimony of many plants may work against them. Most fruits contain a high proportion of water and very little sustenance. In consequence they are easily and swiftly digested, and many fruit-eating birds may void seeds within two minutes of swallowing them.

Plant tissues, of course, are eaten not only by birds but many other animals as well. No plant is without insects that, given the chance, will munch its leaves, sip its sap, masticate its wood, burrow in its fruits and seeds or chew its roots. And such insects then become food for birds.

The little song wrens of South America spend virtually all their time searching for insects on the ground, lifting up fallen leaves and throwing them around to peer beneath. Sometimes they burrow so deeply in the leaf litter that they completely disappear and only pop their heads up above it every now and then, like swimmers taking breath.

The European tree creeper takes not only adult insects such as earwigs and cockroaches, flies and beetles but also their immature forms, caterpillars and grubs, and looks for them in the bark of trees. It starts low down on a trunk, probing the bark with its small curved beak, and clambers upwards, occasionally making little hops and spiralling around the trunk as it goes. It climbs with its feet wide apart and braces itself with the stiff pointed feathers of its tail pressed against the bark. It is so adept and confident a climber that when it extends its search along the branches it, as often as not, clambers along their underside. Having searched one tree, it will then fly off obliquely downwards to another and start a climb all over again.

Woodpeckers hunt for insects which are beyond the reach of the tree creeper, beneath the bark and even in the wood below. The implements they use are not tweezers but chisels. A blow from a woodpecker's bills strikes a tree at about 25 miles an hour and does so with such force that the two mandibles of its beak would fly apart were they not, at that moment, held together by a special locking device. The shock of the blow is so great that if it travelled directly to the brain, the bird would be knocked unconscious. That does not happen because its brain lies above the level of its beak and is cushioned by muscles at the base of the beak which act as shock absorbers.

Having hammered its way into a gallery excavated by an insect larva of some kind, the woodpecker has to extract its prey. This it does with an extraordinarily

Right: a pileated woodpecker,
North America

long tongue. In some species, it is four times as long as the bird's beak. To accommodate it, a sheath extends from the back of the beak over the rear of the bird's skull, along its crest and down over the front of the bird's face. Salivary glands around the sheath at the base of the beak coat it with a glue-like mucus and this, together with the barbed hairs with which it is coated, makes it stick to the soft skin of a beetle grub so that the woodpecker can drag out its victim. This equipment is invisible within the woodpecker's head so we tend to forget that it is there, but it must surely rate as one of the most extraordinary devices for collecting food possessed by any bird.

No woodpecker has managed to reach the Galapagos Islands. Only a few landliving birds have made the six hundred mile ocean crossing from South America and they did not include woodpeckers. But insects did and their larvae now burrow away inside Galapagan trees just as they do on the mainland. The finches however did not allow such a rich resource of insect food to remain untapped. Nor did they wait for evolution to provide them with the chisel beak and the long probing tongue of the woodpecker. They improvised.

A Galapagos woodpecker finch, seeking a meal, hops along the branches of the small shrubby trees, cocking its head to listen for the tiny noises that betray the presence of a beetle larva chewing its way within. Having detected one, the finch makes a hole with its beak, which is scarcely more pointed or powerful than that of a house sparrow. Then it searches for a cactus spine that is the right size to suit its purpose. Holding the spine in its beak, the bird inserts it into the enlarged hole and begins to push it back and forth and from side to side. If it fails to make contact with the grub, it may decide that it needs a rather larger entrance hole. In that case, it does not discard its tool, but either puts it under one of its feet or parks it like a pen in an inkwell, in a hole it may have made earlier, while it tackles a little more carpentry. Eventually it is likely to impale its prey and haul it out on the end of its probe. Then, holding the spine once more under its foot, it removes its meal with its beak as if it were the last morsel on a kebab stick.

This triumph of invention and manipulation entitles the Galapagos woodpecker finch to be called a tool-maker just as much as the human being who first chipped a flint, and it has brought the bird world-wide fame. But more recently another bird has been observed which has, if anything, even greater proficiency as a tool-maker for it manufactures and uses not just one kind of tool but three. It is a species of crow and it lives alongside the near-flightless kagu on the island of New Caledonia in the Pacific. The first tool is a poker, a sharp leaf stem several inches long. With this the bird probes around in the crowns of palms. Here, buried in the accumulation of leaf litter lurk big fleshy grubs. Sometimes the crow manages to impale one, rather in the same way that the Galapagos finch does. But it has also learned that if it does so repeatedly, the grub will become so irritated that it will

bite the spike with its powerful jaws and then hang on with such determination that the crow can pull it out. The second tool, used for much the same purpose, is rather more elaborate. It is a kind of hook. The bird carefully selects a twig with a curved end, breaks it from its branch and takes it back to its perch. There it removes the bark and any leaves it may have and spends several minutes using its beak to exaggerate the curve of the hook on the end. Its third implement is a harpoon, made from the long stiff strap-like leaves of the pandanus. These carry lines of backward-pointing spines down each margin like the teeth of a saw. The crow tears a strip from the edge of one of these leaves, and holding it in its beak with the saw teeth pointing upwards, jabs it at grubs so that the teeth snag in the grub's soft skin.

Insects are wonderfully accomplished in the air, but birds are so much bigger than they are that insects stand little chance against those birds that specialise in hunting them. Swallows pursue them individually, jinking through the air with spasmodic flutters of their wings, opening and closing their forked tails as they execute the most elaborate aerobatics in pursuit of their tiny prey. Swifts are less discriminating. They trawl through the air, holding their beaks wide open. It seems a

Above: a woodpecker finch
probes for a grub, Galapagos

much less energetic way of feeding than the powered pursuits of the swallow and indeed it is. But the rewards are smaller, for whereas a swallow often chases and collects quite large insects such as flies, beetles and even butterflies, a swift feeds predominantly on much smaller gnats and mosquitoes that float more or less passively in the air. As a result, in terms of the amount of food collected compared to the amount of energy expended in doing so, the swallow is almost four times as efficient as the swift.

Nor are insects safe from birds at night. Nightjars pursue large ones such as moths and beetles using the same technique as swallows use during the day. Their beaks are short in length but very wide at the base. The lower mandible also has a special joint half way along its length on either side that allows the tip of the beak to drop even lower. As a consequence, the nightjar's gape is huge. Lines of feathers modified into bristles fringe the beak on both sides. It was once believed that these funnelled insects into the bird's mouth, but now that has been questioned. Some authorities have suggested that they have a defensive function, deflecting insects away from the bird's eyes. Others maintain that they are some kind of sensory device, like a cat's whiskers.

Insects are not without their defences. Like plants, some have powerful stings and others carry poison in their tissues. But some birds have learned how to overcome these too. Bee-eaters, spectacularly-coloured fast-flying birds that live in Africa and Asia, are expert in dealing with even such formidably armed insects as wasps and hornets. A hunting bee-eater stations itself on a perch beside an open space regularly used as a corridor by its prey and waits for one to appear. Bees, wasps and hornets are all vividly striped with yellow and black. It is a signal understood throughout the animal kingdom to mean danger. The bee-eater, however regards it not so much as a warning as an invitation. As soon as such an insect appears, the bird takes off and gives chase. If the insect detects that it is being pursued, and many quickly do, it will dive for cover, perhaps among the leaves of a tree. The bee-eater will not follow it there. But if it does not find safety quickly, it has little chance of escape. The bee-eater quickly overhauls it, seizes it in mid-air and holding it transversely in the tip of its slender beak, takes it back to its perch. There it kills its victim by striking its head sharply against its perch. Then the bird turns its head to the other side and rubs the insect's abdomen against the perch so that its venom is discharged into the air. The insect, dead and disarmed, is then safe to eat.

The North American lubber grasshopper does not have a sting. It is also seemingly defenceless for it lacks wings. But it does have a potent acrid poison in its body and advertises the fact with black and scarlet stripes. Not all young birds are instinctively aware of the meaning of this warning. Sometimes they attempt to eat a lubber but they quickly disgorge it. They have learned their lesson and seldom make a second attempt. The loggerhead shrike, however, has a way of dealing even

with this. Like its European cousin, the great grey shrike, it catches insects of many kinds and stores them by impaling them on the thorns of the tree in which it habitually perches. But, like an experienced cook who knows that some game is much better to eat after it has been allowed to hang for some time, the shrike will only make a meal of a lubber grasshopper after it has remained in its larder for a day or so. By that time, the poison in the grasshopper's body will have degraded and lost its potency. The European great grey shrike even extends its diet to include small birds and little mammals such as voles and mice.

Other animals, intentionally or unintentionally, often help birds in their hunts. Many of us in Britain develop regular partnerships with a robin who unfailingly appears whenever we choose to dig in our gardens, boldly loitering within a yard or so of us, nipping in to pick up a newly exposed worm or grub that it would have had difficulty in excavating for itself. No doubt the alliance is an ancient one that formed when human beings first started to dig in the ground. Before then, robins may well have relied on the grubbings of wild pig, as they still do in many parts of mainland Europe. Nightjars probably established a partnership with wild cattle and goats, collecting moths and beetles disturbed by their hooves. They still flutter around domesticated flocks and herds at night, and do so with such regularity that country folk once believed that the birds were seeking to take their animals' milk and accordingly gave them the wholly inaccurate alternative name of goat-sucker.

Above: a great grey shrike, with
its impaled prey, Spain
Following pages: cattle egrets on buffalo, Kenya

Cattle egrets in Africa follow at the heels of grazing antelope, elephant or rhinoceros, collecting insects kicked up by their hooves that otherwise would have remained safely hidden in the grass and the soil.

Oxpeckers form a relationship that is even more intimate. They are relatives of the starlings and about the same size and they find their prey on the bodies of their partners – big game, such as antelope, giraffe, buffalo and rhino. There is plenty of to be found. Ticks and lice crawl through the hairs of an antelope's skin, drinking its blood. Maggots hatch from eggs laid by flies in fresh wounds and steadily gorge themselves on the damaged tissues. The parasitised animals cannot possibly by themselves rub off all these damaging passengers. Some may be in their nostrils or deep in their ears. So they tolerate the searches of the oxpeckers, allowing them to undertake the most intimate of toiletries.

This association has existed for so long that the oxpeckers have become adapted to it to an extraordinary degree. Their toes and claws are particularly long and sharp, enabling them to hang on to their hosts as they move. Their tails are short and stiff and act as props just as those of a woodpecker do. Their bills are flattened from side to side so that by turning their heads to one side, they can comb through an animal's coat, getting beneath the hairs that lie flat and close to the skin to reach the ticks. They could hardly build their nests on their hosts, but they do the next best thing by picking off their host's hair and taking it away to line their nest holes in nearby trees.

Although it may have been external parasites that initially drew the ancestral oxpeckers to the bodies of big mammals, such insects now form only a minor part of their food. One recent study has discovered that about a quarter of the oxpecker's diet is wax and skin taken from its hosts' ears and a similar proportion is dandruff from the hide. But the birds' favourite food is blood. This they may obtain when they swallow the ticks whose bodies are bloated with it, but they also take it directly, pecking at an animal's wounds to stimulate the flow. It seems that the birds' services to their hosts is considerably outweighed by the damage they do in keeping an animal's wounds open to infection. The flesh and blood of mammals - meat - is, in fact, the most nutritious of all foods. The oxpeckers only nibble at it. Other birds have found ways of consuming it in quantity. But to do that, they need special weapons in order to kill their victims, and special tools with which to execute their butchery.

Left: yellow-billed oxpeckers on a giraffe's neck, Kenya

4

MEAT-EATERS

The kea is a parrot. Its home is New Zealand and although it does not have the flamboyant colours of many of its Australian relatives, it is nonetheless a handsome bird with scarlet beneath its wings of muted green. Like all of its family, it has a powerful, sharply hooked beak which it uses to crack nuts, mash fruit, and nibble any odd bits and pieces lying around that it considers might perhaps be edible. It is a hardy bird, better able to tolerate cold than many other parrots, and lives happily in high mountain valleys, clambering around snow-covered boulders and gliding across the face of the great precipices.

One small community of keas haunts a desolate valley in New Zealand's South Island, where the mountains run down steeply into the sea. There are colonies of sooty shearwaters here too. You never see these birds during the day, for they are fishing out at sea, but at night, they return to the nest holes they have dug for themselves in the turf among the boulders. The local people call the shearwaters 'mutton-birds' because the young squabs, by the time they are four months old, have fed so well on the semi-digested fish brought back to them by their parents that they are full of fat and weigh a couple of pounds. The first human settlers to the island used to harvest them in great numbers. And so do the keas.

A kea with a fancy for such a meal, stalks through the warren of shearwater nest holes, bending down every now and then, head cocked, to listen. Within the burrows, shearwater chicks crouch silently in the darkness. But sometimes they call. Maybe they mistake the tread of the kea for that of their parent, returning with food. If one does make some sound, the kea reacts swiftly. It starts to dig. Using its beak like a mattock, it tears away the turf around the burrow's entrance. It reaches inside, but the shearwater's burrow can be a long one. The kea may have to dig for several minutes more, scattering the brown earth behind it, before it leans inside once again. It is now within reach of the chick. The young mutton-bird is not entirely defenceless. Its stomach may still be full of the fish oil fed to it by its parents and it squirts it directly in the kea's face. The kea is unlikely to be deterred.

Doggedly, it reaches into the hole again until at last it is able to grab the chick by its neck or leg and drag it out. The young bird is still without feathers and covered only in down. It looks as fat and as vulnerable as a plucked duck and it squeals in distress. But not for long. The beak that was so effective a mattock now becomes a billhook and rips the young shearwater to pieces. The kea may be a parrot, but it has become a hunter.

Eating meat brings one great benefit. It is much more nutritious, pound for pound, than nuts or fruit so a meat-eater need only spend a small part of its day feeding. Nevertheless, collecting meat can be a major problem. Animals do not yield their flesh as easily as plants surrender their fruits and seeds or even their leaves. Keas probably took to hunting relatively recently in their evolutionary history and even now do not rely on it as the principle source of their food. Their hearing, on which they rely for finding their victims, is no more acute than that of most bird; and their beaks, so effective for cracking seeds, just happen to be also good for digging. Specialist meat-eaters, if they are to be successful, must have more refined senses with which to locate their prey than the kea possesses and more deadly weapons with which to despatch it.

Above: a kea, South Island, New Zealand

The great grey owl that lives in the dark cold coniferous forests of the north has the most sensitive of ears. The range of frequencies it can hear is much the same as ours, but its sensitivity over that range is very much greater and it can detect sounds so faint that they are completely inaudible to us. It has two large ruffs on either side of its face, formed from fine hair-like feathers that collect the sound waves and deflect them into its ears, rather as an ear-trumpet does. The ruffs also shield each ear from the other, so that sound coming from the right side is received almost entirely by the right ear. Furthermore, its two ears are not symmetrically placed on its head. One is higher than the other. The owl is thus able to listen in super-stereo and can locate the source of any sound with great accuracy, even if it cannot see the creature that is making it.

That is an essential talent during the winter, when the snow lies thick on the ground. The owl, perched on the branch of a pine tree, twenty feet up, suddenly detects the sound for which it has been waiting. Fifty yards away and two feet beneath the surface of the snow, a vole is nibbling a frozen leaf and making just the faintest rustle. The owl swivels its head to face the direction from which the rustle comes and to locate it more accurately. Once it has fixed its position, it topples forward. Its downward glide is virtually silent for the primary feathers on its wings have comb-like fringes which deaden the sound of air passing across them. The little mammal beneath the surface of the snow hears nothing to warn it of what is to happen. Equally importantly, the owl can continue to hear the faint sounds that first attracted it. For a few moments, it hovers just above the snow, staring downwards. It is not focusing its eyes, for there is still nothing to be seen. It is focusing its ears. It pounces. There is a flurry of snow as its wings strike it and the owl reaches through it with its legs outstretched. With a powerful flap, it lifts itself into the air again carrying a still struggling vole transfixed in its talons. Silently, with a few slow flaps, it returns to its perch. It lifts the little furry body up to its beak and swallows it whole, in the way that all owls take their meals.

A few voles are quite enough to sustain it for some time. The rest of the night and the following day, it can sit immobile, well nourished and quietly resting, digesting its meal. Inside its gizzard, the vole's body is macerated. Meat and guts are separated from fur and bones. These inedible components are then rolled by the gizzard's muscular walls into a pellet which, in due course, the owl will regurgitate and let fall to the ground.

The great grey owl has to depend primarily on its hearing, not only because the snow conceals its prey but because for half of the year, the Arctic forests are in partial or complete darkness. Farther south, owls can use their eyes for hunting. All birds must have excellent eye-sight for it is essential in controlling their flight. But flying hunters have a special need for acute vision. The bigger the eye, the more light it can gather and owls, which so often hunt at twilight or by the light of the

Right: a great grey owl locating its prey, Yellowstone, USA

moon, have very big eyes indeed. The scops owl, which hunts in European and Asiatic forests, has such enormous eyes that they cannot revolve in their sockets. If it wants to look to one side, it has to turn its whole head. Owls are particularly good at that. They have a neck joint that allows them to turn right round and stare directly behind them if they wish. Some can even turn their heads beyond that and revolve through 270 degrees.

The retina, the screen at the back of the eye on which the lens of the eye casts its image, contains two different kinds of sensitive elements. One set, the rods, register shape; the other, the cones, register colour. Animals that sleep at night and are active mostly during the daylight hours have a mixture of rods and cones in their retinas. But colour is largely invisible at night and nocturnal creatures have eyes with retinas packed almost entirely with rods. So we can be reasonably sure that many owls perceive the world largely in monochrome but are able to do so in light so dim that other creatures would not be able to see at all. At such low light levels, it is movement more than shape that catches the attention. A mouse, crouched motionless on the forest floor, may be almost invisible to an owl. But if it moves, then it will immediately reveal itself. And since a mouse *must* move if it is to gather its food, it cannot be safe from the owls that hunt in the stillness of the night-time forest.

Birds that hunt during the day sacrifice low-light perception of shape for sensitivity to colour and have both rods and cones in their retinas. The acuity of their vision is nonetheless extraordinary for their retinas, in the case of some eagles, are physically even bigger than ours; and the rods in them are more densely packed. The retina of a human eye contains some two hundred thousand rods. That of a buzzard contains about a million. If our eyesight is particularly good, we may be able to spot a rabbit flicking its ears from a hundred yards away. A buzzard is able to do so from a distance of two miles.

Some species are able to see over an even wider colour spectrum than we can. We have three different kinds of cones in our retinas and we detect different shades by combining signals from all three in rather the same way that printers reproduce different tints by using three different inks in varying proportions. Birds, on the other hand, have five or six different kinds of cones and some, certainly, can perceive ultra-violet light. The kestrel is one and it has recently been discovered how this might have value for the bird. Voles are one of its main prey. The little animals run along regular tracks, collecting their vegetarian meals, and mark these paths with little squirts of urine. This helps them find their way and conveys messages of ownership and sexual availability. But it also reveals their paths to a kestrel, for urine reflects ultra-violet light. So the bird, hovering above, knows just where it should look to spot a give-away movement.

*Right: a crowned eagle feeding on
a vervet monkey, South Africa*

The weapons used by hunting birds are simple, unique – and lethal. Many other kinds of birds can and will, on occasion, kill other animals. Jays and magpies will eat the chicks of other birds and herring gulls catch mice when foraging on refuse tips. But such omnivorous birds kill with blows of their beaks. The specialist hunters – owls and eagles, hawks and falcons – kill with their talons. These are long, sharp and so deeply curved that when they stab into an animal's body, only a deliberate action by their owner will release their terrible grip. Three talons point forwards; a fourth, longer than the others, extends backwards. A hawk, clutching the head of its victim, will often slide that rear claw into the back of the skull and into the brain. It is sometimes called the killer claw.

The talons of the eagles of tropical forests are so massive and so sharp that they can stab right through the body of a monkey. The African crowned eagle catches monkeys that weigh as much and sometimes more than itself. As it dives on them, it swings its legs and pelvis forward, and strikes its victims with such force that the blow may kill them outright. If it does not, then the wounds inflicted by the huge talons stabbing into their internal organs are sufficient to do so. The harpy eagle of South America is even said to take sloths, which are the size of a sheepdog, grabbing them as they hang somnolently from the branches of rain forest trees. Indeed, eagles are said to be the only creatures that can reach these bizarre animals to prey on them. In the Galapagos Islands, an endemic hawk, closely related to the buzzards, hunts marine iguanas, seizing the black, two-foot-long reptiles in its talons

and eventually, often after a long struggle, killing them by tearing them apart with its beak.

Each kind of hunter has its own technique for deploying its weapons. A golden eagle, hunting rabbits, is unlikely to be successful if it dives directly from the sky on to its target. The rabbit will see it coming and scamper to safety in its hole before the eagle can reach it. So instead, an eagle will soar at altitude over its hunting territory, scanning the ground beneath. Having spotted a grazing rabbit, it will flap away and gently lose height. Then it flies back swiftly, only a few feet above the ground, and comes out of nowhere to pounce before the rabbit has time to get back to its burrow. The sparrowhawk also attacks from low-level. It will fly fast along one side of a hedge and then suddenly vault over to catch small birds unawares perched on the other side. The merlin takes small birds in the air. Larks are a favourite prey and the merlin tries to catch them by dropping down from above. A lark, therefore, if it becomes aware of the merlin's presence, seeks safety by climbing. It rises nearly vertically, frantically pumping its wings and seemingly standing on its tail. The merlin gives chase, circling up after it. As long as the lark remains uppermost, it is safe, but if the merlin seems to be climbing faster, then the lark will have to change tactics. Suddenly, it closes its wings and drops like a stone. If it manages to pitch into a bush or thick grass, then as long as it keeps completely motionless it will be safe.

The peregrine, closely related to the merlin but significantly larger, uses the same tactic of diving on its victim from high altitude but does so with even greater speed. Split second timing and perfect flight control are essential. Somehow, the peregrine must keep its eyes fixed on its victim, to judge its velocity and course, and adjust its own trajectory accordingly. What happens in the final stages is far too fast for the human eye to follow. It can only be seen with the help of slow-motion film. In the last few moments, the peregrine actually slows down slightly. A fraction of a second before impact, it brings its legs and talons forward and strikes. The rear talon, the killer claw, is the main contact. It rips along the back of the target and up to its neck, breaking its spine and sending it tumbling from the sky.

Sometimes, not surprisingly, the peregrine overshoots. Then, almost unbelievably, it may pull out of its dive and strike its prey from the rear on the way up again. This sudden change of direction must produce such a gravitational pull on its body that a human fighter pilot might well wonder why the peregrine does not black out. The skills required to hunt in this way are so finely tuned that even a misjudgement by a microsecond will cause failure and the fact is that a hunting peregrine only succeeds in less than one in ten of its attacks.

Above: a peregrine snatches a pigeon
from a pool

In many hunting species, there is a considerable difference in size between the sexes. The male sparrowhawk is almost half the size of the female. This is because the female, as the breeding season approaches, has to accumulate bodily reserves from which to make her eggs. Her ovaries swell and her weight increases by as much as 13 per cent. This increased weight makes it difficult for her to achieve the speed and agility that she needs in order to catch sparrows and other small birds. So there is a division of labour between a pair. She, with her added bulk and size, mounts guard over the nest and defends it against predators. He – small, swift and agile – does the bulk of the hunting and brings back his catch to her and, in due course, to their chicks. Later, when the chicks are bigger with increasing appetites, she may help to feed them by snatching pigeon squabs from their nests, something that he is too small to manage without great difficulty.

Falcons, hawks and eagles, unlike owls, do not swallow their prey whole. Often it is far too big for them to do so, for a peregrine may take a grouse and an eagle kill a hare. It is now that their beaks come into action. Falcons have a distinctive little tooth half way along the cutting edge of their upper beak which they use to give their prey the coup de grace, jabbing it between the vertebrae of their victim's neck, but for the most part, hunting birds use their beak for butchery. Holding the body of their prey on to the perch with their talons, they rip it apart, taking the meat and favoured parts of the entrails such as the liver, and discarding the bones, feathers and fur. They do not, therefore, need to regurgitate pellets of inedible material to the same extent that owls do.

Some hunting birds even dare to tackle terrestrial animals on the ground. The European short-toed eagle takes on particularly dangerous prey – snakes. It goes into battle with its wings outstretched and uses them in the same way as a bull-fighter uses his cape – to deceive its opponent about where its body is and where an attack might come from. The snake, a grass snake or even a viper, strikes repeatedly and accurately enough at what it sees, but all it manages to hit are feathers and any venom discharged is wasted. It might, with luck touch the eagle's legs, but even if it did, it would inflict little damage, for the eagle's bony legs are protected by particularly stout scales and have few blood vessels in them that would carry the snake's poison to vital parts of the bird's body. The eagle prances back and forth, keeping just out of range of the snake's striking head until, in a split second, it sees its chance and grabs one of the snake's coils in its talons. The snake is now at close quarters and still has enough freedom of movement to stab the eagle's body with its fangs. The bird must act fast with its other weapon, its beak. With one stabbing bite, it severs the vertebrae just behind the snake's skull. The battle is over.

If the eagle has chicks, it will need to get its capture back to the nest. That is indeed a problem. A long snake, dangling from its claws, would not only be unwieldy in the air but would be an open invitation to others to try to steal it. The solution is

simple. The snake is swallowed. Inch by inch, the eagle gulps down the still writh-ing coils until only the tip of the tail is left protruding from its beak. Back at the nest, the snake's body takes almost as long to come out as it did to go in. The eagle tries to pull it out with its foot but it cannot get a proper grip on it. One of the chicks tries to help. It grabs the snake's tail and pulls. Out comes the whole snake. The chick seems in no way dismayed by the sight but sets about swallowing a meal that looks to be almost bigger than itself.

Two birds in particular have become so skilled at snake-hunting that they have largely abandoned flight. The seriema, a distant South American relative of the cranes, will pick one up by the tail and kill it by repeatedly beating its head against the ground. The African secretary bird which feeds on all kinds of ground-living creatures from insects to rats makes its kills by stamping on its victims.

All meat-eaters, however, do not do their own killing. They find what they want by waiting beside the hunters' tables. In Africa, marabou storks, bald of head and scabrous of bill, stand impassively beside the kills of lions, waiting for the moment when they may be allowed to pick scraps of meat from the bones; in the Antarctic, sheathbills scuttle through penguin colonies seeking dead chicks and the yolk of

Left above: a secretary bird stamping on grasshoppers, Africa

Above: marabou stork at a lion's kill, Kenya

Left below: a short-toed eagle disgorging a snake, France

broken eggs; in the Falkland Islands, where other food is in extremely short supply even pintail ducks will rip scraps of meat from the carcasses of seals that so often litter the beaches. But the specialist scavengers of the bird world are the vultures.

Just as nectar feeders in Africa and South America have come to resemble one another, even though they are not closely related, so too have vultures on the two continents. African vultures are specialised relatives of hawks and eagles. South American vultures, on the other hand, are related to storks. Internal differences in their skeleton and musculature make this quite clear. But both have broad, blunt-ended wings that enable them to circle on the currents of warm air that rise above the grassy savannahs; and both have naked heads, for they feed by thrusting their heads inside corpses and feathers would quickly become soiled by blood and guts and become the source of infections.

The bare heads of both may also bring an additional advantage. The temperature at the altitudes where the birds spend so much of their time is very low and the plumage on their bodies is consequently particularly dense so that they remain warm as they soar. Down on the ground, however, it can be very hot indeed. Many vultures cool themselves by squirting urine over their legs which, as it evaporates, cools them – and also, incidentally, leaves them looking as though they have walked through white-wash. But on the ground, the bare skin of their heads may also be of great help in getting rid of unwanted heat.

It has long been a matter of argument as to how vultures, circling high in the sky, are able to detect a corpse lying on the grasslands below. Do they do so by sight or by smell? This apparently innocent question sparked off one of the most spectacular and vitriolic feuds in natural history. It started in 1825 when Charles Waterton, a choleric Yorkshire squire and an ardent naturalist, published an account of his explorations in the forest of Guyana. In it he reported, with no particular claim to originality, that turkey vultures there were attracted to carrion by smell. The book brought him national celebrity. The following year, however, another naturalist from the Americas, Jean-Jacques Audubon, appeared in Britain, bringing with him his spectacular drawings of American birds for which he was

seeking a publisher. He too made a great impact on British society. He wore buck-skin jackets, had hair down to his shoulders and told thrilling stories of adventures on the American frontier. Two such outspoken and extravagant characters both claiming expertise in the natural history of the New World could hardly fail to cross swords. Audubon picked on Waterton's account of the turkey vulture and gave a lecture 'exploding the opinion generally entertained of its extraordinary power of smelling'. Waterton took up the challenge and published a letter saying that Audubon's claim was 'lamentably false at almost every point'.

So battle was joined. Each side of the argument attracted followers. In America, college professors and lecturers joined together to sign an affidavit supporting Audubon. One supporter conducted an experiment in his back garden in which he presented a vulture with a painting of a dead sheep and a barrowload of fresh offal. The bird attacked the painting and ignored the offal, thus proving to the experi-menter's satisfaction that it relied on visual not olfactory stimuli. Waterton stuck to his guns and declared that Audubon should be horsewhipped for his 'gross exag-gerations and errors in ornithology'.

The controversy continued, almost unbelievably, throughout the rest of the cen-tury. Reading the arguments now, it seems that much of the disagreement stemmed

Above: a turkey vulture soaring
in search of food, Florida

from the fact that the fundamental difference between African and American vultures had not been recognised and that no one had allowed for the possibility that even the closely related black and turkey vultures could have different sensitivities. Audubon's supporters seemed prepared to attribute any sense to any vulture as long as it was not an olfactory one. One man claimed that a vulture was brought down from high in the sky by sound – the buzzing of flies around carrion; another that it was the sight of small flesh-eating rodents converging on a corpse that caught the birds' attention. Yet a third invoked an 'occult sense' which he explained disarmingly was *so* occult that there was no way that an experimenter could detect it.

While naturalists argued, more pragmatic people came to their own conclusions. In the 1930's engineers in California were having trouble with locating leaks on a thirty mile stretch of pipeline carrying gas across rough country. Someone maintained that turkey vultures could not only smell but were particularly attracted by mercaptan, the pungent chemical put into commercial gas supplies which are otherwise odourless and therefore dangerous. The percentage of mercaptan in the pipeline gas was significantly increased and groups of vultures assembled wherever it escaped, showing the engineers exactly where repairs were needed.

Even so, ornithologists did not resolve the matter to their own satisfaction until 1964 when experiments were conducted with turkey vultures that rigorously and painstakingly excluded any other possibility of their being guided by any clue other than smell and demonstrated beyond any doubt that this is how turkey vultures are guided to their meals.

The turkey vulture is, nonetheless, exceptional. Its nostrils are larger and internally more complex than those of its relations and the section of its brain that interprets the signals conveyed from its nostrils is also considerably bigger. The black vulture, closely similar in appearance except that its naked head is black and not scarlet like the turkey vulture, has it seems no sense of smell. Neither have the other two South American species, the king vulture and the condor. Consequently, turkey vultures are nearly always the first to discover carrion. They feed as quickly as they can, but black vultures will soon notice them doing so. They arrive in force, usually greatly outnumbering the turkeys, and are so aggressive that the turkeys are pushed aside. They thrust their long scrawny necks in through the dead animal's mouth and up its anus to rip out the entrails. But neither blacks nor turkeys have the strength to tear a hole in the hide of a large animal such as a cow. Before long, however, king vultures descend on the black scrum. As they land, the smaller vultures defer and allow them to rip the hide. Last of all come the huge condors. They strut towards the corpse shouldering aside smaller rivals. The carcass is torn apart, its entrails spill and the feast becomes a frenzy.

No African vulture has the power of smell to rival that of the turkey vulture. They do indeed rely primarily if not entirely on sight – and not only to spot a minuscule speck lying prostrate on the plains below. They also keep a sharp watch on the behaviour of other birds circling in other thermals. As soon as one tips sideways and starts to descend, the others realise that there is food to be had and glide across, either directly in a long shallow glide or indirectly from the top of one thermal to the bottom of another and up again until at last they come down on their target.

Just as there is an order in which different species feed on a carcass in South America, so there is in Africa. Griffon and white-backed vultures, are usually the first on a corpse. The biggest of the African vultures, the lappet-faced, does not usually arrive until the feast has already started, but it is so large and powerful that the smaller species give way to it. It struts towards the body with exaggerated steps, displaying its power and aggressive temper to lesser breeds and begins to rip at the hide. In fact, it prefers to eat the hide and sinews rather than the entrails. The smallest of the group, the hooded and the Egyptian, wait on the fringes of the feeding scrum, picking off scraps and waiting until the bigger birds have eaten their fill. They will do the final clearing up.

Above left: a soaring condor, Argentina
Below left: condors at a dead horse, Peru

Above: a lappet-faced vulture on a kill, Kenya

But African vultures do not necessarily have the carcass all to themselves. Hyenas also scavenge. They will have kept a sharp eye on the birds in the sky and noticed if a succession of them start to descend and land. The birds can glide down faster than the hyenas can run across the ground so they are likely to get there first. They eat so quickly and gluttonously, stuffing their capacious crops with gobbets of flesh and entrails that much of the carcass may have disappeared before the hyenas get to it. But its flesh may not be entirely lost to the hyenas. The vultures by now have eaten so much that they have real difficulty in getting off the ground. The hyenas chase them and if the birds fail to get into the air, they will disgorge some of their meat. The hyenas stop to grab it – and the vultures, with somewhat lighter loads, manage to fly up laboriously into the lower branches of the nearest tree.

Before long, little will be left of the kill except bones. The hyenas will finish these off, using their heavy grim jaws to crack them open and expose the succulent marrow. But in more temperate parts of Africa and in southern Europe, a bird is able to do even that. The lammergeier, the bearded vulture, positively prefers bones to meat. It pulls a skeleton apart and removes the long limb bones. They contain marrow and that is so rich that their diet, consisting of 70% of marrow bones with a little meat and skin, is calculated to be 15% richer in energy than one which consists entirely of meat.

Above: Rüppell's vultures at a kill, Kenya

The lammergeier does not have a cracking tool that can compare for power with the hyena's jaws. It has another way of solving the problem. It picks up a bone and flies up high above a patch of bare rock. Then it drops it. It will do so as many as fifty times until eventually the bone hits bare rock in just the right way to split it. The lammergeier then picks up the splinters and, holding one vertically in its beak, takes it into its throat like a circus sword-swallower. The bone disappears downwards, for the bird has exceptionally powerful digestive acids in its stomach, and the bottom end is crumbling away, even as the top is entering its beak. It can totally digest a cow's vertebra in two days.

There are not many sources of food left untapped by birds, either on the earth or above it.

Above: a lammergeier swallowing a bone, Pyrenees

5

FISHING FOR A
LIVING

The waters of the earth, either salt or fresh, not only cover most of its surface but in some parts are even richer in food than any equivalent area on land. Not surprisingly, therefore, some of the animals that evolved on land have turned to the waters for their food. A few reptiles have done so – crocodiles and turtles. So have a few mammals – otters, seals and whales. But no group of land-living animals have done so in such numbers and in such variety as the birds.

Birds need few physical modifications to enable them to snatch morsels from the shallow edges of the water. That most brilliant of all British species, the incomparable little turquoise-blue kingfisher, sits on its regular perch above an English stream, short of tail but armed with the long dagger of a beak. At the sight of a small fish in the water below, it will flash into action. If its perch is a low one, only a few feet above the surface of the water, it will fly upwards to gain height to give itself room to gain speed on its dive. Then down it comes, increasing its momentum with a few flickering wing-beats. With wings extended but folded back tightly against its body, it plunges into the river. Its target may be a minnow or a stickle-back. Even if it is as much as three feet below the surface, the kingfisher will be able to reach it and seize it with its beak. Beating its wings below water to help it rise, the little bird shoots up through the surface and flies off to a perch. There it kills its victim by thrashing its head against some hard object and, with one gulp, swallows it. The whole act is a display of extraordinary skill, performed without hesitation or fumble, and can be completed in a few seconds.

The African pygmy kingfisher, in size and coloration, is remarkably like the European species. Watch it feeding and you will see a very similar performance. But there is one crucial difference. This bird lives, not on a river bank but in the rainforest and it catches not small fish but insects. In fact this is more typical of

Right: a kingfisher seizes a fish, Portugal

kingfishers as a group, for some two-thirds of the family live away from water. In Australia, they grow large and take land-living prey of some considerable size – the kookaburra captures lizards and snakes. So the kingfisher's superbly controlled accurate dive seems to have been developed originally to pluck prey from the ground. The birds that pioneered diving for fish had to learn how to offset their aim to allow for the way that light bends as it enters or leaves water; and some developed the muscular skills needed to hang poised in the air while taking that aim. They achieved this without the radical skeletal modifications evolved by the hummingbird. Their hovering skills are not, it must be said, as great for they cannot maintain a position in mid-air with the perfect steadiness of a hummingbird. Most need the assistance of a little wind in order to hang on trembling outstretched wings. But one of them, the African pied kingfisher, is able to hover even in still air. And that brings it a major advantage over other kingfishers, for it is no longer tied to a perch but can take up a stance wherever the fishing seems most promising.

Other birds walk around the water's edge looking for their prey. The wry-bill, a small plover, searches among the pebbles of New Zealand rivers. Uniquely, it has a beak which is bent sideways – and always, for some reason, to the right. By cocking its head to the left, the bird can slip this remarkable asymmetric instrument beneath a heavy pebble, open its mandibles and scrape off fish eggs, insect larvae or anything else edible that might be clinging there. One wonders why it, alone among the birds of the world, should have developed such a curious technique. Maybe the reason is connected with the fact that, until recently, New Zealand had no land predators. Danger, for the wry-bill, can only come from the air, in the shape of a hawk. Feeding with its head bent not downwards but to one side enables its left eye to keep scanning the sky for trouble.

Herons also fish along the margins of the water, moving with extreme stealth, freezing into total immobility the moment a slight flash in the water gives a hint of a meal. The little black heron of Africa and Madagascar shades its eyes while searching. As it stands with water half way up its legs glaring intently at the surface, it brings its wings forward to form an umbrella over its head. And there it

crouches, for minutes at a time. Sometimes, without changing the position of its wings, it sticks its head up from between them and looks around, just to see what is going on elsewhere, and then draws it back in again to resume its watch on the water. The obvious explanation for this performance is that it is shading its eyes from the glare of the sun just as we do with our hands. Maybe it is simply cutting out reflections from the surface of the water. Maybe the explanation has nothing to do with optics and more to do with the behaviour of fish. They often seek a shaded part of a stream where they are less visible to fishermen. Maybe the heron, by creating a pool of shade, is persuading them to swim within range.

Herons are certainly ingenious and inventive fishermen. In Japan, green-backed herons attract their prey with bait. A few years ago, some were living beside a lake in a public park where people come to feed ornamental fish. One particularly observant or inventive heron watched them and started doing the same thing. It picked up a piece of bread or some other edible fragment that it did not eat but fish did, took it down to the edge of the lake and flicked it on to the surface of the water. A fish came up to take it – and was itself taken by the heron. The habit spread.

Above: a black heron fishing, Africa

Then the technique became more sophisticated. Human fishermen have long known that fish are inquisitive. An object need not necessarily be edible to attract them. A bright piece of metal spinning in the wake of a boat will trigger a snapping reaction. So will a small brightly-coloured piece of fluff. Birds too, have been exploiting this tendency for a very long time. The little egret, which has black legs but bright yellow feet, will stand on one foot and shake the other in the surface of the water like a limp yellow glove, and attract fish that way. The green-backed heron in Japan, having had its success with pieces of bread, has recently started to use little feathers for the same purpose. And that too works.

Skimmers also benefit from the urge fish have to investigate the unusual. The skimmer has the oddest of beaks. Its lower mandible is almost twice as long as the upper. The birds, as big as gulls with black and white plumage, sit in flocks on sandbanks. In Africa, they live on rivers. In North America, the species seems to prefer coastal lagoons. But they only live beside waters that are still and regularly free from waves. When they decide to fish, they take off and then, flying so low over the surface of the water that their wing tips almost touch it, they open their beaks so that the elongated lower mandible cuts a furrow in the water. If that mandible touches something hard, it will snap upwards, closing the beak and catching whatever triggered it. With luck, this may be a fish rather than some solid floating object. Either way, the shock on the beak of the swiftly flying bird must be

Above: a black skimmer feeding, Florida

considerable, but the skimmer has particularly strong muscles in its head and neck which act as shock absorbers. They are seldom successful on this first traverse. But having reached the end of their stretch of river, they turn round and fly back over the course they have just covered. The ripples are still spreading and often by then have attracted fish up to the surface to investigate. So the skimmers' second run is often more fruitful than the first.

Entering the water in order to find food clearly presents greater problems than making brief dips with the beak. Because birds are warm-blooded, they run the risk of getting seriously chilled. Swimming mammals, such as whales and seals, guard against this by developing swathes of fat around their bodies which insulates them very effectively. Birds, if they are still to fly, do not have the option of such a weighty expedient, but they have their own unique and very effective alternative – their plumage. The feathers that enable them to fly also trap air and keep them warm on land. They will do the same thing under water, provided they are able to retain that air.

The European dipper relies on that. It is a small brown bird about the size of a thrush with a white bib that lives on the banks of swiftly flowing rivers and streams. It has a particularly large preen gland on its rump and takes special care of its feathers, anointing them with water-repelling oil. When it decides to feed, it walks down to the water and with no break in its pace disappears below the

Above: a dipper feeding underwater, Europe

surface. In spite of the flow of the water, it is able to walk along the stream-bed, turning over pebbles with its bill, probing beneath bigger ones, searching for the larvae of caddis flies and other insects. If it needs to swim, it does so with rapid beats of its short, rounded wings. To protect its eyes and to enable it to see underwater, it has the equivalent of a diver's goggles, transparent membranes that it can draw across its eyes. And thanks to its well-oiled feathers, it shows no sign of chilling even though it may remain below the surface for up to ten seconds.

The disadvantage of using air as insulation is that it is buoyant. The dipper has to hold on tight with its toes to prevent itself from bobbing up to the surface. The anhinga, which lives in somewhat warmer parts of the world than the dipper, does not have that problem for it has dispensed with such insulation. It has no preen gland and its feathers actually absorb water, so when it enters a river it quickly becomes soaked to the skin. Its buoyancy is so low as a result that when it swims its body is almost submerged with only its long thin neck easily visible. This, swaying from side to side as the bird swims, looks rather like a snake and for this reason the anhinga is often called a 'snake bird'. Such low buoyancy brings a special reward. The anhinga is able to walk along the stream bed and stalk fish with the stealth, the

sudden frozen postures and the swift attack that herons use out of water. It may even use an underwater version of the black heron's trick of opening its wings to attract small fish to their shade. After a successful hunt, it emerges with the fish impaled on its dagger-like beak. With a vigorous shake of its head, it detaches the fish from its bill. Then it throws it into the air, catches it and swallows it head first. But drenched plumage, like a wet bathing costume, needs to be dried quickly if its wearer is not to get chilled. So when the anhinga emerges from fishing it hangs its feathers out to dry by sitting on its perch with its wings outstretched.

Snakebirds and dippers, herons and kingfishers live beside water. Ducks live on it. Because they are so familiar, we tend to forget how remarkably specialised they have become to the aquatic life. They rest, feed, court and mate on its surface. To do all that, they need efficient paddles and their toes accordingly have become connected with webs of skin. Just as the most effective position for a ship's propellers is at its stern, so a duck's legs are placed far back on its body. As a result, while a duck is as admirably agile on the water as it is in the air, the best it can manage on land is an ungainly waddle.

Some ducks, including mallard, teal and pintail, feed on the surface of the water. Their bills are lined with rubbery comb-like plates, one along the edge of the upper mandible and two others running along the edge of the lower mandible between which the top one fits. They feed by drawing water in and out through these plates and filtering off small crustaceans and bits of vegetable matter. Others swim down

Left: an anhinga catching a sunfish, North America

Above: an anhinga with a speared fish

to rootle in the mud. The pochard collect roots and bits of vegetation. Eiders collect molluscs. The tufted duck is omnivorous taking crustaceans, insects or seeds according to the season and the locality. The goosander and the smew not only eat shrimps and worms but are also very adept at catching fish which they chase actively under water. Their bodies are more slim and streamlined than most other ducks and their mandibles are finely notched, like saws so that they are able to catch and hold small slippery fish.

Divers, or loons as they are called in North America, eat nothing but fish. Their webbed feet, placed at the very far end of their body, give them great speed under water. With their wings clasped tightly against their body so that their streamlining is near perfect, they pursue their quarry and often manage to out-manoeuvre them in their own medium. The great northern diver regularly swims down to a depth of seventy feet and stays below the surface for as long as a minute or more. Out of water, however, it is as clumsy as it is agile in it. The bird can barely walk. It flops down on its breast and can only move forwards by holding its feet together and making short awkward hops.

Lakes are not, in the long run, permanent features of the landscape. The rivers that supply them with water also fill them with mud and sand. Over the decades, their waters shallow. Reeds growing around their margins are supplanted by bushes and small trees. Slowly, lakes turn into swamps. Then new kinds of fish and new kinds of fishing birds come to them.

Left: a tufted duck searching for food underwater

Above: a goosander fishing, northern Europe

In hotter parts of the world, the lakes may each year evaporate altogether. If that happens, the birds that feed in them will be able to fly elsewhere, but the fish are trapped. They must have ways of surviving the drought. The African lungfish, at the beginning of the dry season, takes the precaution of burrowing into the mud at the bottom, wrapping its tail around its head and secreting slime. If and when the swamp is baked dry, it will be able to survive in a state of suspended animation, obtaining all the oxygen it requires by absorbing it through a pair of finger-shaped projections from its gut. The shoebill stork prevents many of them getting to this stage. It is a truly formidable bird. Its powerful massive beak is so big that the bird is also known, justifiably, as the whalebill. It stands four feet tall. As it wades slowly through the water, clogged with vegetation, it holds its huge bill vertically downwards so that it can focus both its eyes on the water. If it spots a lungfish lurking on the mud at the bottom or feels it with its feet, it lurches forward with its whole body and plunges its bill into the water. Using its wings as arms, it pushes itself upright again with a foot-long lungfish wriggling in its great beak.

As the season progresses, the lakes continue to shallow. Fish that swam in comparative safety become increasingly vulnerable and birds travel from far and wide to claim a share in the banquet that is in prospect. Now those with long legs have conditions that best suit them. Striding through the muddy waters, yellow-billed storks, herons, and egrets stab and thrust at their wriggling prey. It is not just fish that are there for the taking. There are frogs, snails and other freshwater molluscs.

Left: a shoebill stork taking a lungfish,   – 131 –   *Above: yellow-billed storks, Kenya*
Uganda   *Following pages, pelicans and storks*
                                                               on a shallowing lake, Kenya

These are the favoured diet of the openbill stork. It has a curved upper mandible so that its beak, when it is closed, has a gap in the middle. It does not, however, use that like a nut-cracker for breaking snail shells. Instead, it first pins down the snail with its upper mandible and then uses the sharp tip of the lower mandible like a knife to sever the muscle which attaches the snail's body to its shell. Once that is done, the bird deftly extracts its meal, leaving the snail shell unbroken.

As the waters continue to shrink, so all the small creatures that once flourished in the deep water at the bottom of the lake come within reach of the flocks of birds, searching the muddy shallows. Spoonbills filter out tiny fish and shrimps with side-ways sweeps of their beaks. Often they work together, advancing in a line abreast so that if small creatures disturbed by the advancing feet escape one beak they are caught by the next alongside. They are even capable of swallowing quite large fish. Ibis probe deeper into the mud with their long downward-curved bills and plovers pick up the swarming insects.

Such conditions occur, not just once a year but twice every twenty four hours along the margins of the world's seas. Every time the tide retreats, the salty waters leave behind a spread of food that is so rich and so regularly renewed that a whole range of birds specialise in feeding here and nowhere else. In such a competitive community, there is no room for the generalist who is reasonably successful at collecting many different kinds of food. Here a species, if it is not to starve, has to be more swift and efficient at reaping a particular crop than any other around. So each kind of bird here has come to favour one kind of food and starts to collect it just as soon as it is made available by the retreating sea.

Dunlins and sanderlings frequent sandy beaches, pattering along the edge of the water. As each wave swills back, they follow it, probing and pecking at any morsel that it has stranded; and then as the next wavelet floods in, they run back again up the shore. On more pebbly beaches, it is turnstones that pace along the water's edge, flicking over small stones with their short, wedge-shaped uptilted bills, picking off little shrimps and other crustaceans.

As the tide falls back still further, it may expose a flat of mud. Oystercatchers regard this as their territory. As the first edge of the mudflat is exposed, they squabble and shriek at one another, jostling to find a place to feed, but as the falling tide reveals more and more, so they are able to spread out and feed without impeding one another. There are two kinds of food to be gathered from the mud – worms and shelled molluscs. Some families of oystercatchers trot swiftly across the mud, stabbing at the head of a worm whenever they see one. They have to move fast if they are to catch a worm unawares before it has time to disappear down its burrow. Other oystercatcher families feed in a much slower way, stalking over the flats and probing deeply with their beaks to haul up a mussel or some other mollusc. They lay this on the surface of the mud and hammer it with their beaks until the shell smashes and they can reach the succulent flesh.

Young oystercatchers quickly learn how to take worms. A wriggling head is an obvious target and a swift and accurate strike brings an immediately gratifying mouthful. Within six or seven weeks, the youngsters belonging to these families are

Left: a white spoonbill feeding, Austria — 135 — *Above: sanderlings gathering food brought in by waves*

catching worms for themselves. The skills needed for collecting mussels take longer to acquire. Adults demonstrate the technique to their young and patiently continue to help them feed in this manner for as much as a year until their offspring get the knack. There are, however, two schools of thought among oystercatchers as to the best way to extract a mussel. Smashing the shell is one method, but if you have a particularly sharp bill then it is possible to push it between the two halves of the mussel shell and cut the muscle holding them together. This is more difficult to do but, once the skill has been mastered, it demands less effort. However, a bird taught to be a shell-smasher cannot easily become a shell-opener because using a beak as a hammer blunts it and it may take two weeks to convert the tool needed for one way of feeding into that which is required for the other. So not only have oystercatchers, as a species, their own special diet, but different clans within the species have their own traditional ways of preparing it.

On muddier shorelines, when the water is only an inch or two deep, food can be gathered by dabbling. The shelduck does that by walking slowly forwards, swinging its head from side to side. The avocet also finds prey here. It is more selective than the shelduck and holds its up-tilted beak slightly open as it scythes it to and fro, feeling for small worms and other invertebrates.

Above: an oystercatcher demonstrating its feeding technique to its chick, North America

Above: an avocet, England

The deeper waters immediately beyond the tidal zone are among the richest areas of the sea. Nutrients washed down from the land by the rivers promote a rich growth of floating algae, the phytoplankton. This is food for small fish which in their turn become the food of bigger ones. And birds from the coast fly out to gather them all.

Gannets and boobies use the same technique as the kingfisher does in freshwater. They dive. In the early morning, parties of boobies leave their roosts on the rocks of the sea cliffs that they have whitewashed by their droppings, and set out to survey the surface of the sea. They travel, often in parties of a dozen or so, flying in line astern and close to the surface of the waves. If they come across an isolated fish, one of them may fly up high and dive down into the waves to seize it, but their main targets are shoals of thousands of small fish, often no more than a couple of inches long and brilliantly silver.

Swimming in a shoal is a form of defence for the fish. An individual has a better chance of surviving an attack by a predator if it is swimming alongside a thousand others than travelling by itself, lonely, isolated and vulnerable. But on the other hand, the whole conglomeration of fish is a very obvious and tempting target for any hunter that discovers it. Big predatory fish lunge into it with snapping jaws and drive the shoal as a whole towards the surface. The mass of the silver bodies creates a pale green patch on the surface of the blue sea. That is the sign for which the boobies are searching. One after the other, they dive on it. Other boobies spot the action from afar. The feeding frenzy begins.

Within a few minutes, a great whirling cloud of birds forms above the shoal. As they begin their dives they pull back their wings and take the shape of an arrowhead, thus reducing air resistance but retaining their ability to steer as they take aim. At the last moment, within a foot or so of the water, they fully extend their wings backwards and hit the surface at 60 miles an hour. At such a speed, the impact with the water must be very violent, but the boobies have a network of airsacs just below the skin in the front of the body which absorbs much of the shock. Nor is water driven into their nostrils for these are permanently blocked and the birds breathe instead through the corner of the mouth. Their dive may carry them down several metres below the surface. They snap at the fish, usually on the way up again and normally swallow it underwater. Back on the surface, they join a raft of others that have completed their dives and rest for a few moments.

Other birds continue to arrive in droves. Presumably individuals as much as a mile or so away may be able to see the commotion and hasten to the spot to join in, but parties from all over the ocean arrive in such numbers so swiftly from all directions that it sometimes seems as though they have another sense that tells them what is going on. The shoal of fish, harried from below and attacked from above, shifts to and fro in the sea. As it comes nearer the surface, paling the water,

Right: a blue-footed booby diving on a fish, Galapagos

increasing numbers of birds are spurred into frenzied action and they dive into the sea with the speed and rapidity of bullets from a machine gun. Each strike raises a plume of white water a foot high. The feeding may continue for half an hour or more until eventually the shoal is either dispersed or manages to get past the predators circling below and descends into deeper water where at least it is safe from attack from the air.

Among the birds that join these feeding frenzies along the Pacific and Atlantic coasts of North and South America, are brown pelicans. They are among the heaviest of all diving birds. They too target individual fish, from thirty feet above the sea. Their dives are spectacular, and they draw back their wings in the same sort of way that boobies and gannets do, but they cannot rival those birds in the depth they achieve. Their bodies are too big and buoyant, so though they hit the water with a considerable splash they cannot reach any fish, even with the tip of their long beak, that is more than three feet down. When they surface, their baggy bills are full of water as well as fish. To get rid of the water, as they must do before they swallow their catch, they have to open their bills slightly. Other birds not so skilled in diving as the pelicans, such as gulls and noddy terns, will be waiting for that moment. Some may even perch on a pelican's head, knowing that eventually the pelican will have to open its bill. And when it does, they try to snatch a fish.

Another species of pelican, the American white pelican, also patrols the coastal waters of the southern United States. It fishes in teams. A party of a dozen or so will alight on the surface of the water and take up a horseshoe formation. In line abreast, they swim slowly forward, periodically opening their wings and plunging their bills into water with that perfect unison in which pelicans seem to delight, whether they are flying or fishing. Fish are driven ahead of them until eventually, the two arms of the horseshoe close and the birds, again with impeccable timing, all dip their heads together and scoop up the encircled shoal.

Neither plunge-divers nor scoopers swim underwater in the sea to any significant degree. But the auks, a family of sea birds which includes guillemots, puffins and razorbills, do actively swim to considerable depths. They propel themselves, not with their feet as loons, grebes and some ducks do in fresh water but with their wings, beating them in the same sort of way as they do in air. But water is very dense and it is only possible to beat wings underwater if they are short and stubby. Wings with such a shape, however, are not very effective in air and have to be beaten very rapidly indeed if they are to keep a bird airborne. Guillemots, consequently, are very poor flyers. To rise from the water, they have to whirr their wings

at great speed and even when they level off, their flight seems little better than panic-stricken.

Penguins have given up the compromise. They have abdicated entirely from the aerial life. Their reward is that they are now incomparably the best swimmers among birds. They are so skilled, they can even outswim many fish. The biggest of all, the emperor penguin can stay underwater for as long as fifteen minutes, swim down to seventeen hundred feet and when foraging reach speeds of over ten miles an hour.

We tend to think of penguins as birds that stand around on icefloes and swim in the near freezing waters of the Antarctic, but several species live in warm waters around the coasts of Australia and South Africa and one lives actually on the equator, in the Galapagos Islands. So in terms of geographical spread, the typical penguin does not live in the extreme cold. There is also a good reason to think that the ancestral penguins were relatively small birds. They must certainly have been capable of flight and no doubt combined that with swimming, as guillemots do today. But flipper-shaped wings fail altogether in air if a bird weighs more than about a kilo. So it is likely that the first penguin was closer in size to the little species that lives on the equator in the Galapagos than the four foot tall emperor of the Antarctic.

Some species of penguins spend as much as 85% of their lives in the water and now have lost several of the adaptations that enabled their ancestors to colonise the skies. Their bones are no longer hollow and lightweight but solid and heavy, so a

Above: a Galapagos penguin fishing
Following pages: chinstrap penguins on an iceberg

penguin is not buoyant and can remain below water with ease. The feathers, once broad with filaments that hooked together to form a vane, have also greatly changed their character. Chilling is of particular concern for a bird that spends most of its time in water even if it does live in a relatively warm part of the world, and penguin plumage has become almost furry. The feathers do not grow in tracts between areas of almost naked skin, as those of most birds do, but are uniformly spread and very dense all over the body. Each feather is short and stiff and has a second short shaft covered with very thin downy filaments that together form an undercoat. On land, where temperatures can fall very much lower than they do underwater, the birds can hold these stiff feathers slightly away from the body so that they trap a layer of air next to the skin. In water, they lie flat and become a coat so dense and well anointed with preen oil that water never penetrates it. Since penguins have abandoned flying, they no longer need to keep their weight down and some of the Antarctic species also have a layer of fat two to three centimetres thick, immediately beneath their skin, which provides added insulation.

The penguin tail is no longer used as a rudder as it is among flying birds, and has become reduced to a stump. Steering is done with the more robust webbed feet that are placed on either side of the tail and so far back on the body that even the lame hopping of the loon is impossible for a penguin. They have two options for getting around on land. They can fall forward on to their bellies and toboggan over smooth ground or snow by kicking out with their feet; or, as they usually do, they can stand vertically, using their stumpy tail as a prop when stationary and plod around upright, flippers clasped to their sides, with that impassive solemnity that makes them, in our eyes, so endearing.

The big penguins of the far south make very long voyages in search of their food. The king penguin has been seen as much as a hundred miles away from its breeding colonies; the emperor, though it does not venture beyond the cold waters of the Antarctic, may travel six hundred miles on a single feeding trip. But some birds fly very much farther across the face of the open ocean. Storm petrels, little birds scarcely bigger than swallows, feed on floating particles so small that they can hardly be seen by the human eye. To pick them from the surface of the water, they face into the wind and hold their trembling wings outstretched to keep themselves aloft while trailing their long delicate legs in the water to prevent themselves being blown backwards. In this position, they hover like ballet dancers on points, sometimes making pattering runs, sometimes little hops and two-footed skips as they peck at the plankton in front of them. From a distance it looks as though they are tip-toeing across the surface and it is said that this matching of St Peter's miraculous if short-lived accomplishment gave them their name.

The distribution of plankton is very patchy but petrels do not rely simply on chance to find it. Their family, which includes shearwaters and albatross, have

tube-shaped nostrils at the base of the beak and, unlike most other birds except the turkey vulture, they have an excellent sense of smell. Just as scent guides individuals back to their own nest burrows at night during the breeding season, so it helps them find food out on the open ocean. When fish or shrimps feed on plankton the tiny plants release a chemical substance and it has recently been discovered that even the faintest whiff of it will attract petrels.

The widest ranging of all ocean-going birds are the albatrosses. They do not dive like the boobies, nor patter across the surface like the petrels. They swoop down and sit on the water and from there catch bigger creatures such as squid which rise at night. They also take any carrion they can find. Their superb gliding skills enable them to soar over the waves with minimum effort and they can remain in the air for months on end, sleeping on the wing. They may fly right round the globe, circumnavigating the Antarctic continent by exploiting the prevailing winds and travelling for a thousand miles on a single food-gathering journey.

But even albatrosses have been unable to break their link with the land. Their need to nest shackles them to it; and their chicks' demand for food compels them to return to it repeatedly. The chick of the biggest of them, the wandering albatross, remains dependent upon its parents for twelve months, so in spite of their ocean-

Above: Elliot's storm petrels
feeding on plankton, Galapagos

going capabilities, adult albatrosses have to return to land throughout their lives and when feeding their young may do so every few days.

One bird has managed to break this tie to a land-bound chick. It has not yet developed a way of laying its eggs at sea; nor has it succeeded in doing so in the air, as it was once believed that a female bird of paradise could do by depositing it on the back of a perpetually flying mate. But once the chick has hatched, the parent does not return to land again that season. Instead, the chick takes to the sea. To that extent, this species must be counted the most truly oceanic of all birds. It is the ancient murrelet.

Murrelets are small relatives of the birds that are known in North America as murres and in Europe as guillemots. It is called 'ancient' because, unlike any other member of its family, it has grey on its shoulders like a shawl worn by the elderly. It nests on islands around the northern rim of the Pacific. Small numbers are found in Japan and Kamchatka but its main nesting areas are on the islands off the western coast of North America.

Frederick Island, a small fragment of land facing the open ocean just off the coast of the Queen Charlotte's Islands in British Columbia, is the site of one of its largest colonies. If you land there in the last week of May or the beginning of June, you will see a few shore birds, oystercatchers, gulls and – significantly – bald eagles and peregrines. When you walk inland, you enter a tall dark forest of giant conifers – hemlock, cedar and pine. The rainfall here is so heavy that the whole of the forest floor is cloaked by thick luxuriant moss. It drapes the rocks, the peaty soil, and the tangle of fallen trunks, softening and concealing all angularities as a heavy fall of snow will do. But here there are even fewer birds to be seen. You may hear the faint high-pitched whistle of a winter wren or occasionally the harsh croak of a crow or a raven. But around you, hidden in small underground burrows, in the soil beneath the logs or at the base of the trees, and in the crevices between the boulders, are at least 80,000 ancient murrelets brooding their eggs or guarding their young.

They have good reason to hide. The peregrines and the eagles, even in this relatively dense forest, have no difficulty in pouncing on a bird out in the open. An adult therefore only makes the journey to or from its nest burrow under the cover of darkness. Frederick Island, however, is so far north that the summer nights are brief and seldom very dark. As a consequence both peregrines and eagles are able to hunt for most of the twenty-four hours of the day. They sit in the twilight on tree stumps or low branches, waiting. The murrelets returning from feeding at sea, come in high and at speed. Their short flipper-wings, that made them so proficient in water, are so inefficient aerodynamically that the birds have to travel fast if they are not to stall. As result, they frequently crash into the branches of the tree above their nest-holes and tumble to the ground. They are not very agile on the ground either. With their legs right at the rear of their body, they cannot stand sufficiently

upright to be able to run without a desperate flapping of their wings. A pair of pere-grines nesting in this forest may catch as many as a thousand murrelets in a single season.

Murrelet parents take turns in brooding the eggs, and change over every three days. It takes some thirty days to complete incubation. Between five and ten per-cent of the breeding population is lost every year. Nor are the adults entirely safe underground. There are deer-mice on the island and particularly large ones. Mur-relet eggs are probably safe from them, for they have such smooth shells that it would be difficult for a mouse to get a purchase on them with its teeth. A newly hatched chick, however, is certainly vulnerable and the parent murrelets never leave any droppings at the mouth of the nest hole or within. Rodents hunt by smell.

At last, the eggs hatch and two little fluffy black and white chicks, with a small white egg tooth on their beaks, huddle together in the nest chamber. The eggs from which they emerged were particularly large and their stomachs still contain a con-siderable amount of yolk. This sustains them for the first day or so, but it also has to provide them with sufficient energy to tackle the extraordinary marathon that now lies ahead of them.

After two nights in the nest – or only one if the egg hatched early in the morning – an hour or so after sunset, the parent birds emerge from the nest hole and begin calling to their chicks with continuous chirps. The chicks come out running. They

Above: an ancient murrelet,
northern Pacific

tumble over the sheer sides of boulders, they sprint across the beds of green moss, they push their way under logs but never once do their legs stop moving. If you pick one up, its little legs continue pedalling at the same frantic pace as if it were a clockwork toy. Their parents move ahead of them encouraging and guiding them with repeated calls, leading them down to the beach.

It may take as much as ten minutes for them to get out of the forest. All that time, they are vulnerable to attack from mice, peregrines and eagles. They emerge on to the beach where they are even more exposed and vulnerable. Their parents are now floating many yards out from the shore, but they are still calling. The chicks rush into the sea, their legs still moving so rapidly that they rise above the surface of the water and virtually hydroplane. Parents and chicks recognise one another's calls. United at last and swimming alongside one another, the adults give their valiant offspring their first meal – a little regurgitated fish. By the following morning, parents and young are out of sight of land. For the next six or seven weeks, they will swim together until the young are big enough and strong enough to collect fish for themselves.

Wandering albatross may make longer journeys across the face of the ocean; emperor penguins may dive deeper into its depths; but the ancient murrelet has come as near as any bird to severing its links with the land.

6

SIGNALS AND SONGS

ew animals, other than ourselves, use their voices as eloquently as birds. Ours are created by the larynx, the lump high in our throats that we call Adam's apple. Birds have a larynx too, but it does not produce sound. It serves instead as a valve which prevents food and water from entering a bird's windpipe and getting into its lungs. Birds' voices come from a different structure that lies much deeper within their body, one that is possessed by no other creatures, called the syrinx.

This box-shaped organ, strengthened around the outside with hoops of cartilage, lies at the bottom end of the bird's windpipe where that divides into the two tubes which lead to its pair of lungs. Within the syrinx, each of these tubes can be closed completely or partially by a pair of fleshy lips. When the bird contracts its lungs, a jet of air is blown through each pair, creating a musical sound. Muscles within the syrinx enable a bird to vibrate each pair of lips independently and so vary the pitch and quality of the note that comes from each.

The windpipe up which the notes then travel also modifies them. As with the pipe of an organ, the longer the windpipe, the deeper the sounds it emits as the air in it resonates. Cranes, which produce trombone-like calls, have windpipes that are so long they curl into loops alongside the keel of their breastbones and in some species actually pass right through it. One of the plainer birds of paradise, the trumpet-bird, has one that runs down across its chest just beneath the surface of the skin and forms a loop. As a trumpet-bird gets older so it develops more loops. Only the males have them and only the males produce the organ-like trumpeting calls that give the species its name – though considering the lavish looping of its windpipe, it might have been more appropriate to call it the French horn bird.

Some birds are able at will to shorten their wind-pipes. Like the syrinx, the wind-pipe is encircled by strengthening rings of cartilage. These are connected to one another by muscles which enable the bird to pull the rings closer to one another like the pleats of a closing concertina. As that happens so the resonances from the windpipe rise in pitch. Penguins have a partition running along the length of their windpipe, dividing it into two unequal halves, so when a jackass penguin brays – it can hardly be said to sing – it produces a two-note chord.

By taking very shallow mini-breaths, accurately co-ordinated with the notes of its call, a bird can sing without a noticeable break in the sound for minutes on end, far longer than even the most highly trained human singer is able to do. Each pair of lips within the syrinx may produce its own sound. Canaries, when trilling, create 90% of the sound with the left tube and use the right mainly for breathing. Cardinals, whistling a glissando, start the lower section with one tube and mix through imperceptibly to the other tube for the higher half. The two sounds from the two tubes may also combine and interact with one another to create a quite different sound. This explains how some birds, such as parrots and mynahs, are able to produce eerily close imitations of human speech, even though they lack the lips and variable mobile tongue with which humans make such sounds. All in all, in duration, variety and complexity, no other vocalisations produced by any other animal can match the song of a bird.

– 155 –

Above: singing blackcap, England
Following pages: scarlet ibis breeding colony,
Venezuela

Birds also communicate visually. Their feathers can, with comparatively little cost to the bird, be coloured and patterned. Muscles in the skin enable a bird to fan them out or erect them to display any message they may carry or, if it is a dangerous time to do that, to depress them and conceal it. They can be given different messages at particular times of the year and at different stages in a bird's life, for the feathers have, in any case, to be regularly changed as they become worn by flight.

Some of the colours of feathers come from pigments. The commonest of these is melanin, a pigment that human beings produce in their skin when exposed to the sun. This creates the black of a blackbird's feathers but also, in different varieties and strengths, browns and yellows. Reds and oranges are created by other pigments called carotenoids. These, however, a bird cannot make for itself. It has to obtain them, directly or indirectly, from plants. Flamingos and scarlet ibis derive them from small crustaceans which in turn get them from the blue-green algae on which they feed. Such birds in zoos, deprived of their natural food, will lose their red colour unless they are given carotenoids in some other form. Turacos, handsome pigeon-sized fruit-eaters from African forests, have a pigment based on copper that is found in no other animal and is accordingly called turacin. It is responsible for the dramatic flash of magenta that appears when a turaco opens its wings.

Other colours are created not by chemical pigments but physical structures within the feathers. The blue of a jay's wing feathers is created by microscopic bubbles in the keratin of the feathers which refract the light. So such a feather will not look blue but brown if light shines not on it but through it.

These two sources of colour, chemical and physical, are sometimes combined in the same feather. The green of the wild budgerigar comes from a yellow pigment overlying a blue created by an internal structure. It has thus been comparatively easy for breeders of pet budgerigars to produce strains which lack the pigmented yellow and so expose the structural blue.

The shimmering iridescence of a male mallard's head in spring, the 'eyes' in a peacock's tail, the almost metallic glint on a Himalayan monal pheasant's plumage and the glittering bib of many hummingbirds, all come from a particular modification in the shape of the filaments on the feather. They are flattened and twisted so that their broad side faces outwards. The filaments can only develop in this way if they are detached from one another, so they lack the tiny hooks which in other feathers link these into a continuous vane. Thus iridescence of this kind is never found in flight feathers. Microscopic layers of melanin particles within the filaments split the light in the same way as a film of oil on the surface of water can do and like such a film, the colour varies according to the angle from which you view them.

Birds use both sound and vision, either separately or together, to communicate with their own particular mates, with strangers of their own kind, with birds of quite different species, and even with utterly different animals. They issue battle cries, send warnings, summon mates, declare war and even, on occasion, tell lies.

Above: a Himalayan monal pheasant – 160 –
sunning itself

Communicating with an animal of a fundamentally different kind, which expresses itself in ways that you cannot parallel and which perceives the world in a quite different manner from you, must present huge problems. Nonetheless some birds do so with such success that they rely on that ability to get their food. Honeyguides do. They are small thrush-like birds that are distantly related to woodpeckers. Most species are found in Asia; two live in Africa. They eat a variety of insect food but all have a taste, not so much for honey as for bee grubs and bees' wax. Indeed, they are the only known animals of any kind that can digest wax. They do this, it is believed, with the help of bacteria in their gut but they manage the feat so efficiently that one has been known to subsist on nothing but wax for a month.

Wild bees choose a variety of sites for their nests. Those that live in the Himalayas suspend them beneath overhanging rocks, like stalactites hanging from the ceiling of a cave. There they are unprotected and Asiatic honeyguides have little difficulty in raiding them. As with all members of their family, the feet of these birds resemble those of a woodpecker with two toes pointing forwards and two backwards so that their owners can cling with ease to a vertical surface, whether it is a tree trunk or a bees' comb. African bees, however, favour more protected sites for their nests – holes in trees or deep in crevices between rocks – and to reach those, honeyguides need help.

Above: a scaly-throated honeyguide feeding on a comb of honey, Kenya

The ratel is a kind of badger and a powerful digger. It will eat all kinds of small animals – frogs, mice, termites, fish, and lizards – but it too has a particular passion for honey and bee grubs. At certain times of the year it lives on little else and so earns its alternative name of honey badger. It is a nomadic creature that wanders widely through the bush, searching for food. As it enters the territory of a honey-guide, the bird will fly down to perch in a bush close to it and call with a distinctive chattering cry. The ratel answers with grunts and ambles towards it. The bird flies off and the ratel follows. Every now and then, the bird stops and calls again, flirting its tail which has distinctive white feathers on either side, as it waits for the ratel to catch up. Eventually the bird flutters up to a higher perch and gives a different call. That indicates that a bees' nest is close by, perhaps in a tree hole. When the ratel finds it, it starts excavations.

The bees retaliate. The ratel is by no means impervious to their stings. It turns, sticks its rear into the hole it has created and releases a foul-smelling secretion from glands beneath its tail. Travellers who have disturbed a ratel at work have said that the stench it creates is unendurable. The bees apparently also find it so, for many leave the nest and the animal, now far less troubled by clouds of angry insects buzzing around its head, rips away the wood to enlarge the entrance hole. Eventually it claws out the combs, dripping with rich deep-brown honey and takes them away to eat. The honeyguide now has its chance and flutters down to feast on what is left of the wrecked nest.

The benefits of the partnership to both animals are clear. The bird, flying every day around its territory, knows the ground intimately and becomes aware of every bees' nest it contains just as soon as one is established. The ratel, wandering into that territory perhaps for the first time, lacks that knowledge but has the strength to excavate nests that the bird could never reach unaided. The partnership can only exist, however, because of the bird's skill as a communicator.

When this relationship was first established we cannot know, but the bird has now evolved such a specialised digestive system that it is likely to have been very long ago. Nor do we know whether the ratel was the bird's original partner. It is certainly not its only one today. The Boran people in northern Kenya are only too glad to enlist its help. A skilled Boran honey-hunter knows just how to call to the bird. He whistles, by blowing across a hollow seed, a shell or the thumbs of his clasped hands. The same little procession then takes place, the bird giving directions both with its calls and its behaviour, the man following. And when the prize has at last been claimed by the man, he by tribal tradition will always leave a section of the comb in a prominent place so that the bird is able to take its share.

Mankind and other mammals, however, are more often the enemies of birds than their partners and the messages birds send to their enemies are simpler and more forthright. The sun bittern uses visual signals. It builds its nest in swampy regions of the South American rain forest. When it is sitting, it is difficult to detect for its brown plumage, barred with thin wavy stripes of grey, white and olive, blends in closely with its background. But if you do see it and walk slowly towards it, the bird, most unexpectedly, will fan its tail and spread its wings, revealing on each a

bright chestnut patch, edged on the upper side with black and heightened by a sur-round of glowing gold. The two patches glare at you like a pair of huge eyes. If you stand your ground, the bird will rise and stalk towards you with wings out-stretched. The effect is so startling and the bird behaves with such confidence that anyone unfamiliar with it might be forgiven for thinking that the creature in front of them was actually dangerous. That, doubtless, is the intended message. It is not, of course, true. The bird is telling a lie.

Owl chicks send similar signals and are similarly mendacious. Climb up to look at a barn owl sitting on its nest on a ledge high in a farm building. The bird opens its wings but holds them low. It fluffs out its body feathers, with its beak held wide open. Every ten seconds or so, it looks directly downwards and shakes its head. This is a most surprising thing to do, for it involves the bird taking its eyes off the intruder and thus, presumably, making itself vulnerable to attack. Nonetheless, the effect, like that of the sun bittern's display, is to alarm an intruder and to give the impression that the bird is more dangerous than it actually is.

The most convincing of all avian lies, however, are not those that say 'I am here and dangerous' but ' I am not here – I am something else.' A great number of birds, like a sitting sun bittern, are camouflaged by their plumage. A woodcock on its nest on the ground in an English woodland, surrounded by the dead brown bracken sits tight and, seemingly, is as invisible to a prowling fox as to a human being. Its cam-ouflage depends on it remaining perfectly motionless. The nancunda night hawk, a kind of nightjar, lives on open grassland in Brazil and spends much of its day rest-ing on the ground. There is, of course, little there to provide cover, but the bird manages to hide by crouching and, helped by its appropriate coloration, disguising itself, it seems, as a pile of cow dung. The resemblance is certainly remarkably close. But there is one difficulty about this explanation. Cows did not graze the Brazilian grasslands until they were introduced from Europe a few centuries ago. The nancunda can hardly have evolved its remarkable imitation in such a short time and there are no other large grazing animals on the plains to provide a model of the bird to imitate. But once there were. A thousand or so years ago, giant sloths and great armadillos the size of small cars trundled across the plains. They are now all extinct, but maybe the nancunda is giving us an accurate picture of what giant sloth droppings looked like when once they littered these grasslands.

The potoo reinforces its impersonations with actions. It is a big bird, a giant re-lation of the nightjars from the American tropics and stands a foot and a half high. It perches, habitually, on a tree stump. Its mottled plumage is much the same col-our as bark so it is not easily noticed at any time. But as you approach, it moves to make its imitation of a tree stump even more perfect. Very slowly, it lowers its tail and presses it against the bark of the stump so the junction between the two is in-visible. Then, equally slowly, it raises its head until its beak is pointing vertically

Right: a potoo in its camouflaging posture,
Brazil

upwards and shuts its eyes. This might seem just as risky an action as the barn owl lowering its head, since the bird has now such confidence in the efficacy of its disguise that it will stay rigid and motionless even if you come to within a yard of it. But the bird can see you, even though its eyelids are closed, for each has two tiny vertical slits in it. They let through enough light into the bird's supersensitive eyes for it to keep a check on what you are doing and how close you are getting. Only at the very last moment will it lose courage and flap away. The effect can be very startling indeed for it is quite likely that until that very moment you were totally unaware that a bird was there at all.

Up in the far north, in the tundra of the Arctic, it is rather more difficult for a bird to disguise itself as part of the landscape, since that changes so radically with the seasons. Nonetheless, the ptarmigan has to do so for it lives there throughout the year. In winter it is almost invisible. Its plumage is as white as the snow across which it wanders looking for the exposed tufts of vegetation that might provide it with a meagre meal. Only its black eyes and beak are visible to the cursory glance and they might easily be mistaken for fragments of rock projecting through the snow.

But as winter retreats, giving way to spring, the snow melts. If the white birds do not change colour, they will become very conspicuous indeed. The long-term solution is to change their feathers for ones of a different colour, but that takes some time and while it is happening, the birds must cluster closer and closer together on the shrinking snow patches. The females complete their moult first. As soon as they

have done so, they fly off into the low bushes of the tundra, confident in the effectiveness of their new brown camouflage, and start to make a nest. The males, however, are still competing among themselves for dominance. A male who has a female already sitting needs to be able to drive his rivals away. It seems that he cannot do this effectively if he is wearing his brown summer plumage. At any rate, the white-clad males start their squabbles at the end of winter and continue them, as the temperature rises, on the dwindling patches of snow. They cannot therefore start their moult until some time after the females, by which time spring is already advanced. Suddenly, their need for disguise becomes urgent. Swift though their moult might be, it still takes three to four weeks and that could be fatally long. The males, however, have found a way of changing from white to brown in a matter of minutes. They fly to a dust-bowl and there they wallow, turning their white feathers brown, matching themselves to their new background.

A bird may also need to communicate with other species of birds. That, perhaps is not as remarkable as communicating with a mammal, but it still a notable achievement. Each species, after all, has its own range of sounds and gestures. But just as human beings in an emergency once used SOS and Mayday signals that were understood by all people, no matter what their native language, so birds also have calls that are universally understood between themselves.

The arrival of a sparrowhawk hunting along a hedgerow is just such an emergency, as far as the hedgerow birds are concerned. The first to spot it sounds the

alarm. Doing so risks drawing attention to itself but the call it utters is short, soft and high-pitched. Such a sound is very difficult to locate. It is often written, accurately, as `seet'. Having made such a call once, the sentinel then keeps quiet and hides in the foliage. All the other hedgerow birds do likewise. Tits, thrushes, and finches each have their own seet calls but all the versions have the same characteristics and all are understood by them all. The mixed community has evolved an international language.

On the floor of the Amazonian rain forest, insect-feeding birds form a similar mixed community and hunt together, rummaging through the leaf litter and overturning bits of bark, to dislodge insects which, as they run for safety, may escape one beak but will be snapped up by another. Two species that join such groups habitually act as sentinels – a shrike tanager which perches in the forest canopy and an antshrike which stays in the understorey. If either sees danger, it will make a seet call and all the birds in the feeding flock will immediately take cover. The sentinels' task requires them to keep looking around and so prevents them from searching in the leaf litter with the same dedication as others in the flock, but their reward is that they are able to collect insects disturbed by others. But they do not always tell the truth. It has been reliably reported that if a particularly large and succulent insect is unearthed by one of the birds in the flock below, a sentinel may quickly utter

Above: an antshrike, Brazil

a seet call, even though no danger has appeared. The flock immediately dashes for cover – and the sentinel swoops down to steal the prize.

A seet is not the only way of sounding the alarm. A very different call is sounded if the danger comes from a hunting mammal on the ground. If it is a cat rather than a hawk that threatens the birds in a hedgerow, the sentinels, instead of remaining unobtrusive and hard to locate when they warn, create a huge hullabaloo, screaming angrily and continuously. Their scolding, of course, warns others and to that extent has the same function as the seet call. But if that was its only message, you might expect that the birds would then fly off to keep out of harm's way. But they do not do that. Instead, they fly towards their enemy, perch in the branches close to it and squall in angry chorus. Unlike the seet call, their shrieks are loud, long and relatively low in pitch, so the position of the callers is immediately obvious.

They are in no real danger. A cat cannot approach and pounce with the speed of a hawk. Its technique is to creep stealthily towards its prey until it is close enough to grab it. Once it has been observed, it cannot win. The screams of the birds tell it that its presence is now known to every bird in the hedgerow and it realises that there is no point in trying to creep up on one of them. At that stage the cat usually abandons the hunt and stalks away with an air which, in a human being, would be construed as one of injured innocence.

Fieldfares, one of the larger members of the thrush family, take even more vigorous action against their enemies. They are very gregarious birds and often nest in colonies. Should a nest-robber and chick-stealer such as a magpie appear near one of their colonies, the first fieldfare to notice it will scream a warning. Its call is quickly taken up by others. Their cries certainly convey their anger to the magpie and are likely to unnerve it, but they also summon the rest of the colony. These are battle cries, summoning the troops. Still screaming, the fieldfares take to the air. Then they begin to dive on the intruder. As each makes its attack, it shrieks until it is within a few feet of its target and then releases a bomb of faeces. Many of these missiles are so well aimed that they actually hit the magpie. Soon its plumage is badly soiled. Sometimes it is so thoroughly plastered that it tumbles to the ground and has to hop away to try to clean itself

If birds of the same species are to have any form of social life, they have to exchange not only alarms and war cries but many other messages. One of the most important concerns identity. An individual male needs to know whether another bird he encounters belongs to the same species as he does. If it is not, then providing it is not an enemy, it may be ignored. But if it is, then the stranger may be a rival for the same meal, the same nest site, the same territory or the same female. Action has to be taken accordingly. So it is that many birds carry patterns on their plumage which declare exactly what kind of bird they are.

Human beings, who also like to be able to identify bird species, have worked out just which colours and patterns on a bird enable them to do that. Any page in a field guide will show you the features they use. A European robin can be recognised immediately by its red breast, a male blackcap by the black feathers on his head. Finches have bodies that are very similar in shape, but each has a different patch of colour on the head – the chaffinch has a grey-blue one, the hawfinch a brown one, the greenfinch a yellowish face with a golden yellow edge to the wings, and the bullfinch red on its cheeks that extends down its breast. Confirmation that these colour patches are the features that birds themselves also use for identification can

be seen if you watch a male bird establishing his territory. The cock robin thrusts out his red chest; the blackcap erects his black crest; and every male finch chooses to demonstrate his proprietorship by alighting on a prominent perch where his splendid uniform in all its detail can be seen to the best advantage.

The comparison with a uniform is indeed apt, for in the days when wars were fought hand-to-hand and face-to-face it was imperative that a soldier knew immediately whether the man who suddenly appeared beside him was a friend or foe. So opposing armies assumed spectacular uniforms with widely differing colours. It was an arrangement between enemies that suited both sides. But within an army, it is also necessary for individuals to recognise something else about another soldier. Given that the overall character of a stranger's uniform indicates that he is fighting on the same side, is he senior or junior? Does he take orders or will he give them? The insignia that convey that information are not so obvious and require closer inspection. How many stripes does he have on his arm? Is there a red tab on his lapel? And so it is with birds.

A male house sparrow has a black bib, but look closely and you will see that it varies in size. A male great tit has a black stripe running down his chest, but some have a broader one than others. A collared flycatcher has a white patch on his forehead but different males have them in different sizes. In all these instances and many more, the dimensions of the colour patches indicate seniority. The bigger it is, the more senior the bird that carries it. If and when two birds begin to dispute, over a morsel of food or the possession of a territory, the junior bird will usually defer to the senior.

Why do they not make a fight of it? The dimensions of these insignia are partly determined by genetics – a vigorous male is likely to pass on his strength to his offspring – and partly by the way that those offspring are cared for. A well-fed young male house sparrow, strong and lusty, will have a bigger bib than a weakly ill-fed one, and a great tit a wider breast stripe. Young collared flycatchers, as they fledge in their nests in Europe and prepare to migrate south for the first time, do not have that white forehead patch. It only appears when they moult on their wintering grounds in Africa. Its size depends on how well they feed there. So when they return to Europe to breed, they carry accurate indications of their strength. If it came to a physical fight, a well-fed male with a big patch would be likely to beat an opponent who has a smaller forehead patch and is therefore not as strong. So it saves energy all round if the weaker bird concedes straight away.

These badges of rank need not be on permanent display. The male wild turkey, that grandest and most magnificent bird of the North American woodlands, has in addition to his spectacular feathers and the strange hair-like tassel dangling down his chest, a naked neck hung around with swags of wattles. These vary in size. A bird that is not very fit, plagued perhaps by an infection of internal parasites, will

Above left: chaffinch; above right: greenfinch   – 171 –
Below left: hawfinch; below right: bullfinch

not have such big wattles as a bird who is in the peak of condition. And as a bird gets older, so his wattles grow bigger. But they can also quickly change colour. Usually they are a purplish red, but if a male disputes with another, perhaps over access to females, then they will rapidly flush bright scarlet. So do those of his rival. As the two contestants square up to one another, the difference in the size of their wattles becomes very apparent. The male with the lesser ones, daunted by the sight in front of him, rapidly backs down – and signals his submission by a rush of blood away from his wattles that, within seconds, fade from bright red to pale pink.

Visual signals, however, have a great limitation. They can only be transmitted and received over a relatively short range. To broadcast messages to a wider audience, sound is much more effective. Some birds create it mechanically. Woodpeckers do so by adapting the hammering action of their bills, which they normally use to excavate their food, to beat out characteristic and meaningful tattoos. As drums they select trees or branches that are particularly resonant – a hollow trunk or a dead branch. Some exploit exciting new instruments that have only recently come their way – a corrugated iron roof or a metal stove-pipe. Each species, indeed each sex, drums in its own characteristic way. The most diagnostic feature, it seems, is the frequency with which each tattoo is repeated. The speed with which the blows are delivered is also important. They are mostly so rapid that it is impossible for the human ear to count them. Only analyses of recordings can do that.

They show that the greater spotted woodpecker hammers out twenty blows a second. By contrast, the Magellanic woodpecker, which lives in the southern beech forests in Tierra del Fuego, the most southerly tip of South America, has a very simple call. It consists of only two blows, but the interval between them is important. Imitated correctly – and they are so close to one another that you will need two sticks to do so rather than one – and the spectacular red-headed bird is likely to appear very quickly and beat an angry response on a tree close by, if not on the top of the one you are using yourself.

Several other birds create mechanical sounds, even though none has such a highly evolved and specialised instrument as the woodpecker's beak. The palm cockatoo in Australia improvises. It breaks off a stick and holding it in its beak, beats it against a hollow tree. Others use their feathers – the African flappet lark by clapping its wings together as it falls from high in the sky down to within twenty feet of the ground; the lyre-tailed honeyguide by diving so that air rushes through slots in its wings to produce a whistle; the European snipe by 'drumming', pitching headlong downwards with its tail fanned and the outermost feather on each side sticking out independently so that the wind rushes past and produces this distinctive noise.

But of course the instrument that most birds use to proclaim their identity is their voice. The distance their calls travel and the degree to which they become

Left: a palm cockatoo drumming beside its nest hole, northern Australia

– 175 –

Above: a bare-throated bell-bird producing its penetrating call, Brazil

distorted is greatly affected, like any other sounds, by the character of the environment. Trees impede them and reduce the distance that they will travel. The leaves, particularly if they are hard and shiny like so many of those in a tropical rain forest, reflect the sound back and forth, so that a complex call becomes slurred and confused. Accordingly, birds that live in the canopy of a dense rain forest use calls that are usually very simple in structure, repeated over and over again in case distortions have prevented accurate reception, and are very often piercingly loud. The bellbird lives in the South American forest. It is little bigger than a jay but it has a voice of the most extraordinary penetration. It perches in the very top of the trees so that its calls are subject to the minimum of interference and will travel farther, and produces a simple two-note call that it repeats over and over again throughout the day. The consequences can be maddening to a traveller as he trudges wearily through the forest, dripping with sweat, and moves from the territory of one calling bellbird to another. Those out of sympathy with birds sometimes call the bellbird the fever-bird.

Low-pitched sounds travel better than higher ones through such congested environments. The small size of the bellbird prevents it from producing deep sounds, but the bittern has the stature to do so. It lives in European reed-beds and produces a deep resonant booming noise which probably gives it the birds' long-distance record for sound transmission. It can be heard up to three miles away across the fenlands.

In one or two celebrated cases, the call is the only way by which, at a distance, two species can be distinguished from one another. Two small warblers visit English gardens during the summer. They have a faint yellowish stripe across the top of the eye and an equally pale flush of yellow on their white breast. If you look very hard, you may be able to detect that one of them has paler legs than the other. But as soon as they sing, the difference between them becomes plain. One produces a stream of liquid musical notes falling in a melodious cascade. That is the willow warbler. The other, the one with darker almost black legs, has a repetitious two-note song, one that is simple enough to convert into words and use as its name – chiffchaff. This pairing was first noted and made famous over two hundred years ago by that most acute observer and greatest of English clergymen-naturalist, the Reverend Gilbert White. There is another pair of look-alikes in the New World that are less famous but equally difficult to separate except vocally – the alder flycatcher and the willow flycatcher. The first lives in the more northerly part of the continent, the second in the south and west. Their ranges overlap where the northern forests give way to the grasslands of the prairies and there the bird-watcher's only hope of distinguishing between the two is by their calls which the field guides transcribe as *way-bee-o* in the first case and *fitz-bew* in the second.

Like visual signals, songs proclaim not only a bird's species but its identity as an individual. Robins living in an English woodland stay in the same territory all the

– 176 –

year round and both male and female vocalise. A male – and to a lesser extent a female – makes a regular circuit of his territory and sings from specific perches. When he stops, he will cock his head and listen. Usually his song stimulates an answering call from across his frontier in a neighbouring territory. He knows all his neighbours individually and recognises them by their song. If the response is one he recognises, then he will continue on his round unperturbed. But if you play to him a recording of the song of another male from another wood, he is likely to react in a very different manner. He repeats his song with greater vehemence and fluffs out his red breast aggressively, preparing himself for a territorial dispute with the stranger. If there really is an intruder, perhaps a young male who is trying to establish a territory between two existing ones, or one attempting to take over a territory made vacant by death, then there may be a vocal duel during which the two contestants size up one another. Other things being equal, a sitting tenant will manage to hold his ground. If the newcomer is merely taking over a recently vacated territory, then the duel will serve to get the two neighbours acquainted before they settle down to accepting the new situation. Only if the established bird is weakly or elderly will the newcomer be likely to establish a completely new territory alongside for himself.

A bird's vocal repertory is partly inherited and partly learned from its parents. That is easily shown experimentally. Chaffinch chicks reared in silence make sounds that are just recognisable as simple impoverished versions of a typical chaffinch call. It is only if they can hear the sounds their parents make that they will vocalise in the way that other adult chaffinches do. Individual birds, however, each have their own individual songs. Since these influence the songs of their young, it follows that over many generations a local population of chaffinches, or any other song bird, is likely to develop its own characteristic accent. And so they do. Chaffinches in the north of England sing songs that are recognisably different to the expert ear from those in the south, and blackbirds descended from those that were taken to Australia during the nineteenth century to gladden the ears of European settlers with sounds of home, now have very distinct Australian accents.

Saddlebacks are among the most loquacious and melodious of New Zealand's native singing birds. They are the size of starlings, black with a chestnut patch across the back and the upper part of the wings so that when they sit with their wings folded, the patch forms a continuous and conspicuous saddle. The males have small orange-coloured wattles dangling from the base of the beak. Local groups do not merely develop their own accent, they introduce so many new phrases and motifs that even a small group develops its own dialect. Each member of a group occupies a territory in the forest and holds it throughout the year, defending it from intruders by singing from regular perches strung out along its

frontier. As he does so, his nearest neighbour will respond to the challenge by sing-ing in response, as a robin does.

Young male saddlebacks, when they fledge, wander unobtrusively through the forest, keeping low in the trees and often feeding on the ground so as not to chal-lenge any of the established territory holders. As they wander, so they come to rec-ognise all the different dialects being sung for miles around. Usually they spend most of their time some distance from where they hatched in areas where their par-ent's dialect is not used. They are on the lookout for widows.

If a male vanishes, the young bachelors detect his absence extraordinarily swiftly and start to sing. One was observed to do so within ten minutes of the own-er's disappearance. But the song a bachelor uses is not that sung by his father. In-stead, he uses the dialect of the area he happens to be in, for this is the version that the widowed female has become accustomed to and which she prefers. If she ac-cepts him, he will take over the territory. As a result of this regular behaviour, in-breeding among saddlebacks is kept to a minimum.

Above: a saddleback calling,   – 179 –
New Zealand

In those cases where a pair of birds hold a territory jointly and the year round, the two may share the task of defending it with song. The to-whit to-whoo call of a tawny owl is one of the characteristic sounds of the English countryside at night, the more noticeable in winter since that is a time when many of the summer visitors have left and many of the residents have fallen silent. The call, with those of the cuckoo and the chiffchaff, is one of the few that is simple enough to be accurately conveyed by written words. But it is more complex than it may seem. It is produced not by a single bird but by a pair. The female is responsible for the first section – to-whit. Then the male, often sitting quite far away, will call oo-oo, his timing being so perfect that the two calls together form a single statement.

Many birds which live together for several years use this method of keeping their relationship firm. African barbets sing their duets sitting close beside one another on a branch. Their performance is timed to within a fraction of a second so that when one bird stops, the other takes up the song with the greatest precision. You might think that such mutual understanding would take a great deal of practice, but should a male from such a partnership be absent, even temporarily, a young male from the family will take his place and perform a similar version of the song, which the pair will very quickly synchronise.

There is a distinction to be made between a call and a song, even though in some circumstances one may grade into another. Calls, which include the seet and other alarm signals, are usually short, simple and uttered by both males and females. They are not learned but genetically ingrained. Songs, on the other hand, are usually much longer, more complex and produced only by males. Although male turkeys and cockerels sing their own celebrated arias, song is largely restricted to passerine birds, that great section of the bird kingdom that contains all the small perching birds.

In the more temperate parts of the world, in Europe and north America, male passerines give their most glorious performances in spring, joining together in a great chorus at dawn. The woodlands and the hedgerows echo and throb to the songs of their avian inhabitants. So many different species contribute that it can be difficult to disentangle one from the other. This leads one to wonder why so many should be singing at the same time. Would it not be more effective if the choristers spaced their songs by singing at different times during the day? Perhaps it is that dawn suits them all particularly well as a time for singing for then, at first light, it is still too dark to go hunting for insects, and still too cold for the insects themselves to be moving around and making themselves easily visible and catchable. Maybe it is because there is seldom much wind at dawn in summer to distort the calls and diminish their range.

It is also the case that sound travels particularly well in the early morning. The rising sun may warm the air before it reaches the land, so that a layer of cold air

many feet deep forms between the ground and the warming skies above. One of the characteristics of sound-waves is that, rather than cross the boundary between cold and warm air, they tend to be deflected by it. Sound waves, instead of escaping upwards into the sky, are bent downwards, and funnelled along this invisible aerial conduit. The phenomenon was investigated on the savannahs of Africa, where it was found that the low-frequency calls of elephants travel five times farther at dawn than later in the day. It also occurs in northern woodlands, and even though the effect there is not so marked, it still causes the songs of birds to travel particularly far at this time of day.

But the primary reason that males sing so lustily and splendidly at this season is certain. These songs contain the most important message of their lives. They are directed to females and they say: come be my mate.

Above: a great reed warbler
singing in the dawn chorus, Europe

7

FINDING
PARTNERS

The process of securing a mate may be imperative but it should also be selective; and it is usually the female who makes the selection. First she must assure herself that a prospective partner belongs to the same species as herself. That is necessary in order to avoid wasting time with a partner with whom she cannot have a fruitful union. The question should be easy for her to resolve for a male has already gone to some lengths to identify himself to his rivals and the signals he used for them are likely to serve her equally well. If there is any possibility of confusion, the males usually go out of their way to make the position clear. The blue-footed booby, which breeds on the western tropical coast of South America and on the Galapagos Islands, shares some of its nesting sites with the red-footed booby. The two species are very similar, except, as their names make obvious, in the colour of their feet. So when a blue-footed male wishes to attract a female, he makes sure she knows who he is by dancing, lifting up his ultramarine webbed feet with all the care and exaggerated movement of a man wearing snow-shoes.

Once that possibility of confusion has been cleared out of the way, a female will next need to satisfy herself that her partner is capable of giving her all the help she needs to raise her chicks. A female European wren expects her mate to provide her with a nest and a male may build up to a dozen nests in different sites before he produces one that convinces a female that he will be an adequate partner. African masked weaver birds build in colonies and females look for a male with a well-sited nest in the colony's tree, preferably dangling from the tip of a high branch so that it will be difficult for a marauding snake to reach it. She also requires that a nest should be firmly woven so that her eggs do not fall through its floor, and be strong enough to hold the chicks as they grow and become increasingly heavy. A male weaver bird seeking a mate must therefore become a speculative builder. Having woven a nest,

Right: a blue-footed booby displaying,
Galapagos

he hangs beneath it, fluttering his golden wings and shrieking every time a likely female flies by. If she approves of his work, she will join him. A youngster in his first year may stand little chance against more experienced males who quickly produce much better nests than the somewhat sloppy untidy version he usually makes at his first attempt. He may be rejected so many times that he becomes discouraged. If the green of the freshly gathered woven strips fades to brown, a female will not even bother to inspect it. Why should she waste her time considering the nest of a suitor who has been rejected for so long by so many? After some time without success, he laboriously unravels his construction to make a second attempt on the same precious site. Often, he has learned from the mistakes of his first, but often too he may not manage to attract a female at all during his first breeding season.

Penguins make minimal nests. The gentoo, which also lives in large colonies, uses little more than a scrape on an expanse of bare Antarctic shingle. Nonetheless, both male and female like to have it ringed by little pebbles and a male will pick one up in his beak and show it to a female as an indication of the sort of thing he might provide were she to join him. The end result is more handsome than might be guessed, with all the constituent pebbles being very similar in both size and shape and most neatly arranged. Adelie penguins do much the same. They make their nests in the far south on the continent of Antarctica itself and have similar nest-making rituals, though for them fragments of ice may serve just as well as pebbles.

Many females will require their mates to supply food for them when they are sitting and for the young when they hatch. A male tern starts his courtship by demonstrating his ability to do this by bringing the female a gift of small fish, held cross-wise in his beak. He continues to make such presentations long after he has been accepted as a partner and does so immediately before each copulation. So while his initial offering might have been a ceremonial proposal, subsequent ones become a valuable element in the female's nutrition while she builds up the bodily reserves she needs to produce her eggs.

The female peregrine has similar expectations of her partner. If he is to live up to them, he has to be a highly competent flyer, so in the air space above the nest site that he has claimed, he demonstrates his prowess by a dazzling display of aerobatics. He spirals upwards to a great height and then plunges down at speed. At the bottom of his dive he swoops up again, sometimes rolling rapidly from side to side, sometimes looping the loop with wings half-closed. The female may join him in this demonstration of aerial skill. The two will swoop on one another, sometimes interlocking talons to tumble downwards through the air, sometimes coming so close together that while in mid-flight they touch their breasts or beaks in an aerial kiss. Tropic birds also indulge in these graceful and breathtaking courtship flights, sailing back and forth in tandem, their long filamentous tails undulating behind them and moving in such perfect synchrony that often it is impossible to tell which of them is initiating the movements of their aerial dance.

Left: gentoo penguins sitting on their
nest-scrapes, Falkland Islands   – 185 –   *Above: a male little tern presenting*
a gift to his female, Sweden

Skills in nest-building, fishing or hunting are obvious qualifications for a mate but in many species, a female makes a rather more generalised assessment to assure herself that her partner is in good health and likely to pass on his strength and vigour to any offspring they may together produce. If he is one of those birds that sing during their courtship, the quality of his song will provide an indication of his suitability. Singing, after all, takes a lot of energy. It also exposes a bird to his enemies, something that could be fatal for individuals who are in less than prime condition. Furthermore, a male who can afford to spend a lot of his time singing is clearly either feeding in a notably effective way or the owner of a particularly rich territory. In either case, he is a desirable partner.

She can also deduce his desirability from his plumage, using much the same criteria as the male displays when establishing his rank among his rivals – the brilliance of the bullfinch's scarlet breast, the size of the white patch on the forehead of the collared flycatcher, the assurance of a blackbird's song and the glossiness of his sleek black plumage setting off the intense yellow of his bill.

So in spring, males are at their most colourful and smartest. The females, who are doing the choosing, have less need to be strikingly coloured and indeed it may be better that they are not, for most will have to undertake some if not all of the incubation and when sitting on the nest it will be safer for them to be inconspicuous.

In those species where the sexes are similar or identical, the birds may select their partners by dancing together. All species of cranes do so. They gather in groups of a dozen or more and begin to bow and leap to one another. They flap their wings, bounce into the air and make sudden frantic runs. Sometimes they will pick up a feather or a twig and throw it into the air as though it were a toy.

Above: the aerial display of cinereous harriers, Chile

Right: sarus cranes displaying, northern India
Following pages, Japanese cranes dancing

Grebes perform a long series of ritualised dances on the lakes and rivers where they live. A pair of great crested grebes start their courtship by swimming up to one another, meeting breast to breast and then, in silence, twisting their heads from side to side, flaunting the broad chestnut ruffs around their necks and the long black ear-like tufts on their heads. Other rituals follow as the days pass. The female will swim with her neck stretched forward and her beak close to the water surface, calling as though looking for the male. As she nears him, he makes a shallow dive and emerges in front of her, holding himself upright with his beak pointing downwards. If over the next few days their relationship matures, they begin to perform the weed dance. The partners swim away from each other and then both slowly submerge. After a few seconds, they reappear on the surface, both carrying a small bundle of weeds in their bills. Ceremoniously, they swim towards one another until they are face to face and almost touching. Then they lift up their bodies erect and, beating their feet, rock their heads from side to side with the weeds still dangling from their beaks. The western grebe, which lives in the western United States, has an even more spectacular climax to its ceremonies. Instead of posturing with weeds, a pair will rise up alongside each other with necks stiffly arched and on thrashing feet scoot wildly across the surface of the water like a pair of water-skiers towed by an invisible boat, until suddenly and simultaneously both dive below the surface.

Left: a pair of great crested grebes performing their weed-presentation ritual, England

– 191 –

Above: a pair of western grebe displaying, North America

Male and female puffins develop special and identical decorations for their courtship. During the spring, they grow a horny outer covering to their beaks, handsomely striped with yellow, red and blue. When breeding is over, this particular extravagance is shed, for it may well be an encumbrance when trying to catch fish underwater.

The brightness of the colours assumed at the beginning of the breeding season tells the females of many species that the particular individual they are considering is indeed of the opposite sex, as does his general bearing and, often, his elaborate and confident song. But some birds provide permanent indications of their sex. Male and female sulphur-crested cockatoos have identical plumage, but the male has a darker eye than a female. The male European greater spotted woodpecker has a crimson patch on the back of his neck which the female lacks. The Namaqua dove, whose call is one of the characteristic sounds of an African evening, differs sexually in that the male has a black face and the female does not.

Other males are even more explicit. The African superb sunbird is indeed superb for the male has a maroon belly and an iridescent head; but only the male deserves that name, for the female is dressed in dull yellow and olive. The Australian king parrot male is brilliant scarlet whereas the female is green. The sexes of another Australian parrot, the eclectus, are so different that for many years ornithologists thought they were two different species. Then someone noticed that one

group of emerald green parrots with scarlet underwings and flanks were all males, and another group from the same area, which were entirely crimson except for their violet blue bellies, were all females and put two and two together.

Male frigate birds declare their sexuality with a fleshy adornment that they display only at the breeding season. They are sea birds that wander across the oceans of the tropics and only come to land in order to breed. They choose remote islands – the Galapagos, Raine Island on the Barrier Reef, Aldabra in the Seychelles and Ascension in the middle of the Atlantic among them. The males claim a nest site first and take up residence. Having done so, they do not leave it, for suitable places are few and another male is likely to claim it if it is abandoned for any length of time. So the nest-holder sits tight and from that position does his best to summon a female by inflating a pouch beneath his throat. For several minutes he repeatedly fills his lungs with air and expels it into his throat pouch. Watching him do it has all the suspense of watching someone blow up a balloon at a birthday party. The billowing scarlet pouch gets larger and larger until it is drum-tight and close, surely, to bursting. As if this huge sign were not conspicuous enough, the male also calls to females flying past by stretching out his wings, vibrating them and making loud gobbling noises.

Left: a male eclectus parrot feeding his mate, northern Australia

– 193 –

Above: a male great frigate bird displaying to his mate, Galapagos

Many male birds sprout special plumes to adorn themselves for breeding. Male egrets grow long white filigree feathers down their backs that they can erect in display. Male ducks go through a complete change of costume. The alteration in their appearance can be so extreme that it is difficult to believe that the gaudy creatures, pirouetting, nodding their heads and erecting banners on their flanks, are the same birds as the dull brown ones that were swimming on the lake only a month or so previously. Mandarin and teal, mallards, harlequins and eider, all change into dandies with the most extravagant costumes. A month or two later in mid-summer, when breeding is over, they will moult their brightly coloured nuptial dress and become once more modest and retiring.

As a result of these displays, male and females form pairs. Some birds find a different partner each year; others, such as swans and albatross, mate for life. Some pairs share tasks of nest-building and chick-rearing more or less equally; others will split just as soon as incubation starts. There are polygamous males and also females who take several male partners. But overall, some ninety per cent of bird species are monogamous.

Some of the males in that remaining ten percent, are not only polygamous, but take no part whatever in family life. Their relationship with their females is limited to the few seconds that it takes to copulate with them. For that to happen, the species has to live in a place where food is so abundant and easily gathered that a female can manage to rear her family entirely by herself. But if a female no longer requires her mate to help her build a nest, or defend a territory or provide food, what quality would make her prefer one male to another? Her only concern need be that he is healthy and vigorous and likely to pass on that strength and general fitness to her offspring. Like monogamous birds, she can judge that, to some extent, from the quality of his plumage, and the way he displays it. Many females certainly use such criteria as the basis of their selection.

This has been demonstrated by a series of ingenious experiments involving the long-tailed whydah, a relative of sparrows with black plumage and scarlet shoulder patches that lives in swamps and marshes in east Africa. The males are polygamous, and hold territories in the reed beds in which their several females build their nests. During the breeding season, they develop broad black tail feathers two feet long and display them in fluttering flights above their own patch of reeds. The experimenters cut off half the length of the tail feathers of some males and increased the length of some others by the same amount. They then counted the number of females each group acquired. The males with shortened tails attracted half as many females as those birds with a normal length; and those with super-tails doubled the number of their females.

Breeding only from the few individuals who exhibit a physical characteristic at its most extreme and excluding the great majority that do not reach that standard is, of course, a technique that has long been used by human breeders of livestock. Pigeon fanciers used it to produce an extraordinary variety of descendants of the wild and unremarkable rock pigeon. They bred pure white birds and all black ones, birds with feathers on their feet, with grossly inflated chests, with exaggerated tails like huge fans, with feathers pointing forwards along the back of the neck to form a hood around the head and many more bizarre forms. Those results were obtained within a few decades. Some wild female birds have been selecting their mates on the basis of their appearance for tens if not hundreds of thousands of years The results they produced became so extreme that different breeds became unable physically to mate with others. So, genetically isolated, they eventually became new species.

Nowhere has this process produced more spectacular results than in New Guinea, that immense thousand mile long island that lies north of Australia. Its forests and swamps are particularly favourable environments for birds. Like most in the tropics, they are swarming with insects, rich in fruit-bearing plants, full of potential food of one kind or another. The island also has another advantage for any bird that lives there. It is, in evolutionary terms, a very young land having been formed some ten million years ago by volcanoes erupting from the floor of the Pacific Ocean. As a result, the only mammals that managed to reach it before humanity were a few marsupials from nearby Australia and some bats. There are no monkeys stripping the trees of fruit or leaves, no squirrels collecting their seeds. So birds there have very few competitors for the forest's food and a female can without too much difficulty gather enough for herself and her chicks unaided. Not only that, but the island has no large mammalian predators either – no jungle cats, no weasels, raccoons or foxes. So attracting attention by displays or being encumbered by extravagant plumes does not necessarily put a bird in danger. Male birds of paradise have exploited these opportunities to an astonishing degree.

The bird of paradise family is thought to be distantly related to the crows. One species does indeed look somewhat like one. This is MacGregor's bird of paradise. It lives in the high mountains of western New Guinea where it can get so cold that the rain forest thins out and is replaced first by dwarf conifers and then open grasslands studded with groves of small stunted tree ferns. There is comparatively little to feed on here so, as might be expected, both parents are needed to collect food for their young and MacGregor's bird of paradise is therefore monogamous. The difference between male and female is minimal and the only ornamentation the birds have are yellow wattles around the eyes and similarly coloured patches on the wings. Otherwise they are black and they, perhaps, give us some idea of the appearance of the ancestral birds of paradise that first colonised the vast island.

The rest of the forty odd species, that now live in lusher circumstances at lower altitudes, vary in size from a magpie to a robin. The females and young males all look remarkably similar, as one might expect members of the same family to do. Most have brownish backs and pale breasts that are barred, speckled or in a few instances, plain. The mature males, however, have such varied and extravagant decorations that it is difficult to believe that they could be related to one another. Some have plumes sprouting from their flanks, others from their shoulders, their chin or their forehead. One wears a tiara of six quills each tipped with a black disc, another is bald with the skin of its scalp a piercing blue. And they flaunt these astonishing adornments in as great a variety of ways as it is possible to imagine.

The black sicklebill, the largest of them all, is entirely black but has feather fans, rimmed with iridescent blue on either side of his breast. The species lives in such remote parts of the mountains that until very recently no scientist had seen it in display, so that one could only guess how these fans were exhibited. It turned out that the bird, perching on his display branch at dawn in the middle of his territory, suddenly erects them so that they frame the whole of his head. At the same time he spreads his long black tail and changes from a shape that is recognisably bird-like

into a sinister looming rectangle. Then he sways and tilts sideways until he is almost horizontal. The smallest species, the king bird of paradise is scarlet with creamy white underparts and two quills projecting beyond his tail, each tipped with an iridescent green disc. He displays by first erecting tufts on his chest and then suddenly toppling down so that he hangs upside down from his branch with his wings open and vibrating. He remains there for a few seconds, then he closes them and with muscular movements of his legs swings his body like a pendulum so that his two tail quills thrash from side to side. The King of Saxony's bird of paradise has two feathers that have such an extraordinary structure that when they were first examined by ornithologists, before the bird itself was found, they were thought to be some kind of forgery. They sprout from his forehead and are twice the length of his body. The quills, instead of carrying normal feather barbs, have on one side only a line of small sky-blue platelets which are so hard and shiny they resemble enamel. His favoured display perch is always a thin dangling vine. When he performs, he throws his two extraordinary plumes forward and then, flexing his legs, he kicks repeatedly downwards, like a child trying to make a swing rise higher

Above: a male black sicklebill, motionless
at the climax of his display, New Guinea

and higher until the whole vine is bouncing. As well as wearing a fantastic costume he is also, in effect, a trampoline artist.

Nor do all birds of paradise display in trees. Because of the absence of ground predators, it is possible to perform in safety on the ground. Some do. They clear a dancing stage on the forest floor by snipping off the leaves from the surrounding saplings and removing all twigs and any other litter from the ground to expose the surface of the earth. If you find one of these arenas in the forest, it is easy enough to discover if its creator is still around and active. All you have to do is to throw one or two leaves on to it and hide. If he is about, you will first hear a few agitated calls and then the bird himself will fly down, pick up the leaf in his beak and, with a flick of his head, throw it away to one side.

The performances enacted on these stages are just as varied as those by other species in the trees. The bird with the blue bald head, Wilson's bird of paradise, has a scarlet back, a sulphur-yellow patch on the back of his neck and an iridescent green breast. When he displays, he clings to the vertical stem of a sapling and distends his breast feathers into a shining green shield fanned out at right angles to his body. The bird with the head pennants, the parotia bird of paradise, looks at first sight to be one of the more sober members of the family for, except for a white disc on his forehead, he is dressed entirely in black. His, perhaps, is the most theatrical performance of all. Before he starts, he inspects his dancing stage with great care, removing any fragment of leaf or twig that might have fallen on it since he was last there. Then he makes a series of scuttling runs across the stage, letting out shrill calls. This announces that he is about to give a performance and usually an audience of several drab females will arrive within a few minutes and take up their seats on a horizontal branch to one side. As they settle in, he makes another tour of the stage. It is immaculate, but even so, just to underline the point, he mimes, pretending to pick up non-existent leaves and then casting them aside.

And now he begins his dance. He lifts himself high and holds out his long body feathers so that, with his wings, they form a circular skirt like a crinoline. He erects his breast feathers so that they catch the light and you can see that although they had seemed plain black, they are in fact iridescent and form a glinting shield, part greenish-blue part gold. Facing his audience, he waltzes from side to side. Abruptly, he stops and for a moment is rigid. Then, standing on the spot with feet astride, he begins to twirl his head so that the pennants of his tiara are lost in a blur. Suddenly he leaps into the air, lands on to the back of one of the females in the audience and mates with her.

Female birds of paradise tour all these varied performances and, one must assume, compare each male and select the one who impresses them most. Human observers who wish to interpret the females' judgement in a strictly practical way may say that the athleticism of a male's dance and the splendour of his plumage are

Right above: a male Wilson's bird of paradise in full display, New Guinea
Right below: a male parotia bird of paradise at the climax of his dance, New Guinea

accurate indicators of his vigour and therefore genetic superiority. Others may also conclude that the female bird of paradise must have an aesthetic sense which leads her to prefer a brighter colour to a dull one, a more extravagant decoration to a more modest one.

Another group of New Guinea birds provides evidence on this point. Bowerbirds are found over much the same geographical range as the birds of paradise and were once thought to be closely related to them, though genetic studies have now shown that this is not the case. One species is golden yellow, another part-yellow part-black, and several others in the family are plain brown and only have yellow feathers in their crests. They too indulge in extravagant courtship displays but instead of using their feathers, they do so with collections of brightly coloured treasures – berries, shells, flower petals, even bits of glass and fragments of plastic. These they exhibit in special constructions of twigs, their bowers.

The toothed-billed bowerbird in the rain forest of northern Australia constructs the simplest display. He merely clears an area of the forest floor and then saws off the large leaves of a particular tree with his toothed bill and carefully lays them out

Above: a male 12-wired bird of paradise with his neck ruff erected in display, New Guinea

on his stage with their pale sides uppermost. Others, including the satin bowerbird, build a yard-long avenue flanked on either side by walls of twigs. At each end of this, the male places his collection of shells, bones and berries. This is strange enough, but one other group, the gardener bowerbirds, all but one of which live in the New Guinea forests, build even more elaborate treasuries. MacGregor's gardener selects a slim vertical sapling and builds a tower of twigs around it. This may be ten feet high. Round its base he clears a circular runway, banked with a low outer wall faced with moss. He then summons females with long liquid calls. When one arrives, he picks up a flower in his beak and dances around the runway, peeking around the central column first from one side and then the other. The striped gardener produces a more flattened version of this construction with the top of the tower extended outwards and down to the ground so that it forms a cave with an entrance on one side. On the floor of this cavern he scatters flower petals.

The most complex bower of all, and surely the most remarkable construction made by any bird, is the hut built by the Vogelkop gardener. The biggest I have seen looked like a man-made hut, for it was big enough – just – to crawl into. Two

Above: a male satin bowerbird beside his bower, Northern Australia

Following pages: a male Vogelkop gardener bowerbird adding to his display of blossom within his bower, New Guinea

saplings in the centre supported its conical roof, thatched with the dried stems of orchids. The ground immediately in front of it was planted with a lawn of moss. On this lay the owner's treasures, neatly arranged in piles according to their character – black beetle wing-covers, scarlet berries, shiny black fruits as big as plums, the huge squat acorns of the local oak, fragments of orange fungi and, most luminous of all in the dim forest light, a great heap of orange-coloured dead leaves. There were at least half a dozen other bowers within half a mile or so. But they were not all the same. Other individuals had chosen other coloured objects for their collections. One had clearly become enchanted by the pink flowers of a creeper that must have just come into bloom in the neighbourhood, for a huge pile of its blossom lay to one side within the hut.

The notion that these bowers and their contents play the same part in the lives of bower birds as plumes do for birds of paradise is supported by the fact that among these gardener birds, at least, there is a direct and inverse correlation between the colourfulness of their plumage and the complexity of their bowers. MacGregor's bird, the maypole builder, has a large orange crest which extends almost half way down his back; the creator of caves, the striped gardener has one which is only two thirds as long, fringed and streaked with brown and invisible unless erected; and Vogelkop gardener, the most extravagant builder of all, is a plain brown bird with no crest whatsoever.

It might seem that the bowerbird's way of displaying is much more economic than growing feather adornments as the birds of paradise do. It does not require a male bird to squander his physical reserves on feather finery that has to be renewed annually, nor does it incommode him in flight. On the other hand, watching these birds at their bowers leaves you in no doubt that owning and maintaining such collections is very labour intensive indeed. The Vogelkop bird worked at his bower almost continuously throughout the day, bringing in new treasures and rearranging those that were already there so that they were displayed to their best effect. He was not able to abandon it for any length of time for if he did, one of the males from a nearby bower would certainly fly in and steal choice pieces. Furthermore some species tend their bowers for as long as nine months in the year. So maybe the size and splendour of a bower is indeed a fair measure of the vigour of its owner and therefore his desirability as a mate. But the fact that the bowers varied in their contents suggests that males had found that it was a viable strategy to proffer particular colour arrangements of their own, which must mean that females have aesthetic tastes that vary from one individual to another. Certainly, female bowerbirds tour the bowers, just as female birds of paradise tour the dancing grounds of their males, apparently assessing their relative merits and then mating with the architect of the bower that takes their fancy beside or even within his prize-winning construction.

One group of birds of paradise, the first to become known in the west and still the most famous, simplify the females' task of selection by displaying in groups. The males have bunches of gauzy plumes that sprout from their flanks beneath their wings and extend well beyond their tails. Those of the greater and lesser birds are golden, Count Raggi's bird has scarlet ones. As many as ten glorious males of one of these species will assemble in a tall tree that may have been used for this purpose for many generations. Their display involves flapping their wings, shrieking and fluffing out their gorgeous plumes and they keep doing so intermittently throughout the day. The arrival of a female, however, sends them into an ecstatic frenzy. They lower their heads and erect their plumes over their backs in a fountain of colour. But each bird stays on his own branch of the tree that he uses every day. It is for the female to fly down and join one of them.

Sustained observation of these group displays reveals that nearly all the visiting females mate with the same male. Have they, in fact, surveyed all contenders who look almost identical to our eyes and all, unerringly, chosen the same individual? Or could it be that in such groups, it is the perch that tips the decision between one resplendent male and another? If that is the case, then it could be that the males' displays are directed not so much to the females as to one another. This might create a ranking system in which the most senior male claims the display perch where all the mating occurs.

Above: a Count Raggi's bird of paradise male displaying to a female

Another tree-displaying bird of paradise seems very likely to be operating in this way. Wallace's standardwing lives only on the island of Halmahera. It is the most westerly of all birds of paradise. Its plumage is not perhaps the most ravishingly beautiful but it has to have a fair claim to be the oddest. It has a slender purple cravat on its breast that normally lies close to its flanks and extends right down to its tail. It also has two long white feathers that dangle from the middle of the leading edge of each wing. The males assemble in even greater numbers than Count Raggi's bird. The species is so rare and little known that for a hundred years after its first discovery by science in the middle of the nineteenth century it was never sighted again. Only recently has one of its display trees been rediscovered. It stands in the midst of the forest and when I visited it, there were so many displaying males that some were displaced into the somewhat lower trees growing beside it. There were between thirty and forty of them. They displayed to one another by erecting their cravat horizontally so that it glinted, purple changing to green, and twirling the white standards on their wings. Those on the periphery of this great assembly called attention to themselves by display flights, shooting vertically into the air on rapidly beating wings, floating for a few seconds at the top of their jump with wings rigidly outstretched and then sinking down again to resume their quarrelsome displays with other males. Determining whether or not females were present was not easy for, as in so many of these birds, there is no way of being certain whether a bird without plumes is a female or an immature male. But even allowing that most were females, it seemed hardly credible that any one of them could survey the whole displaying multitude to make a proper judgement between them. Yet here again, all the copulations were taking place on just one branch and, seemingly, with just one male. If it is indeed the perch that determines the females' choice, then the extravagance of the males' plumage is due, not so much to the females' delight in beauty but to the males' assessment of what is most daunting.

Several other families, as well as the birds of paradise, display in this way, the males dancing either in one dazzling group or separately, each on his own stage but sufficiently close together for the females to move easily from one to another to make their comparisons. This phenomenon is known as a lek, a term derived from a word meaning flirtatious play, first applied by Scandinavian ornithologists to ruffs, sandpipers that dance in this way. The male ruffs grow extravagant plumes on their heads and necks which vary considerably in colour, some black some white, some reddish dappled with black. A dozen or so will assemble on open land and there parade while the smaller soberly-dressed females move among them to select their mates.

The buff-breasted sandpiper breeds farther north in Canada. They are more discreetly dressed and dazzle the females by showing off their white underwings. When a female approaches the lek a male will flash open one of his wings. If this

action catches the female's eye, she will pause in front of him waiting for a repeat performance. He may open either wing, but if he becomes particularly excited he suddenly spreads both while several females cluster in front of him, often totally ignoring another male sitting nearby who – no matter how energetically he exposes his armpits – seems not to have what it takes.

Several bird families seem particularly prone to this kind of polygamy. In Europe and Asia there are pheasants and their near-relations the grouse. The peacock belongs to this group and has perhaps the most spectacular adornment of any

Above: a male buff-breasted sandpiper displaying to females, Alaska

bird but there are others that come close to it in extravagance and magnificence. Bulwer's pheasant has an ultramarine wattle hanging on either side of his face. When he becomes excited, the upper half of his wattles above his eyes engorge with blood and rise up as two vertical spikes, while the pair beneath the eyes quadruple in length and extend right down his chest. The great argus pheasant is equipped with immensely elongated wing-feathers which he fans out into a vertical screen. He then peers at the female through the centre of it with one glittering eye. The sage grouse spreads his tail so that the pointed feathers form a spiky sunburst behind him, lowers his folded wings, and inflates a huge pendulous air sac on his chest. As it increases in size this exposes two patches of naked olive-green skin and eventually conceals his head altogether.

Left: a male great argus pheasant at rest, perched in front of a female, South-east Asia

Above: a great argus pheasant male in display

In South America, richly costumed displays are exploited to a spectacular degree by the cotingas. Male umbrella birds have pendulous lappets hanging from their throats, red or black according to the species, and black feathers on their heads that are equally long on all sides and hang down so that they look more like badly fitting wigs than umbrellas. They call with a deep flute-like note from the tops of trees, and females fly from one male to another. The calf-bird, like lekking ruffs, displays in large groups, but in tree-tops. At first sight they seem rather drab with uniform russet-brown bodies, black wings and tail, and a naked head. But they become more dramatic when they display – rearing into a near-vertical position on the branch, everting a pair of small cream-coloured globes on either side of the tail and letting out a quite unexpected call. Their name comes from its sound, but to compare it to a calf's bellow does it an injustice: it certainly has a moaning quality, but with strange electronic overtones. They produce it by inflating a throat sac, which swells so much that the feathers which normally cover it lift away from the sides, exposing the throat sac's walls, which are so transparent that you can see right through them.

The most brilliant of the family is the cock of the rock – the males a startling red or orange, with a feather cockade on the head which extends almost to the tip of the beak. While awaiting the approach of a female, they perch 10 feet or so above

Left above: cock of the rock on their lek, Guyana

Left below: two male calf birds displaying, Brazil

Above: a bare-necked umbrella bird, Costa Rica

the ground in the forest undergrowth, squabbling and squawking among themselves. But when she arrives they all flop down to the ground, each on his own small plot, and there squat and start to make little bouncing hops. The greyish female shows her choice by fluttering down behind one of the males and giving him a sharp peck on his rump. For a few seconds he stays motionless as though he can hardly believe his luck. Then he suddenly turns and jumps on her back for a swift copulation.

By and large, the more elaborate a bird's costume the less complex its vocalisation. It seems that a bird has little need to invest in both. The plumage of the peacock and birds of paradise may be flamboyant, but their calls are harsh and simple. But there is one conspicuous exception to this generalisation. The male Australian lyrebird has long, gauzy tail feathers flanked on either side by a curving feather with its vane richly patterned in brown and cream. The males each have their own display mounds in the southern Australian forest, on which they dance, and the females tour them to make their choice. When the male displays, he bends his tail forward over his back and fans it out in a most spectacular fashion. But at the same time, he sings one of the longest, most melodious and complex of all bird songs. This must mean that the females not only admire visual beauty but have, in addition, a predilection for vocal brilliance and coloratura. The males, in their need to increase the length and variety of their song and outdo their rivals, have become superlative mimics. They include in their cascades of trills, warbles and liquid notes, the songs of almost every bird in the surrounding forest. Even the most inexpert bird-watcher can identify and admire the accuracy of a lyrebird's kookaburra impersonations, but a skilled ornithologist may be able to recognise the songs of over a dozen other birds embedded in the lyrebird's incomparable recitals. Some individuals have territories close to those occupied by human beings and they incorporate the new sounds they hear coming from across their frontiers. So they include in their performances accurate imitations of such things as spot-welding machines, burglar alarms and the camera motor drives. The female lyrebird, it seems, is an aesthetic glutton. Like an over-demanding opera-goer who is only satisfied if her tenor has a slim athletic figure as well as a dazzling voice, she tries to find a mate who is as beautiful to look at as he is to listen to.

Most avian polygamists are male, but in a very few instances it is the female who has multiple mates. The red-necked phalarope, a small wader, nests on the ground on the open Arctic tundra. Being so exposed, its eggs are frequently lost to predators. The females of many species, after such a disaster, will lay again. And so does the phalarope. But she has made the practice a routine one. Having mated with one male, she leaves him to take care of the incubation and moves away to find another male who will do the same thing. In one season, she may have as many as four mates and four families of chicks. It is advantageous for any sitting bird to

Left: a male lyrebird in display, with his tail feathers thrown forward over his back and head, Australia

be well camouflaged so it is not surprising to find that the male phalarope is dully coloured. The female, in contrast, is the one who has the redder neck. She is also bigger and she is the one who will display when initially claiming a nest site.

The American lily-trotter or jacana also nests in dangerous circumstances. In the lakes and ponds where it lives, there are threats from crocodiles, snakes and birds such as purple gallinules that will readily eat eggs or chicks. A female will claim large tracts of this dangerous territory and attract as many as three males, each of whom builds a nest within her domain. She stays with each male for a few days of courtship before she lays a clutch of eggs in his nest but thereafter she may never visit him again. She does, however, maintain watch on the whole of her territory and zealously drives off intruders. Why this reversal of the normal role of the sexes has taken place among jacanas and phalaropes is still an unanswered question.

The complexities of courtship, monogamous or polygamous, are only the preliminaries leading to the crucial act of copulation. For an event of such importance, it is completed in a remarkably short space of time. Indeed, many people seeing it may understandably doubt whether it has happened at all. The male hastily and awkwardly clambers on to the back of the female, she bends her tail to one side, he to the other, the two genital openings are pressed together for a second or so, the male tumbles off and it is all over.

Male birds, with very few exceptions, do not have any organ such as a penis which can be inserted into the female to place the sperm deeply and securely inside her body. Why should this be? Perhaps it was lost, ancestrally, as part of the trend towards reducing weight. Maybe it also has something to do with the fact that maintaining your balance while standing on two legs on a narrow perch with another individual on your back is not an easy thing to do for any length of time. Even on the ground, a lengthy copulation could be dangerous since most birds have only too many terrestrial enemies and may need to take to flight at a moment's notice. So it may be better for the pair not to be physically connected so intimately that separating within a second or so is difficult. Whether that is so or not, sperm is transferred from male to female almost instantaneously by a fully co-operative act. A bird has only a single rear vent, the cloaca, which leads to both the end of the gut and the sex ducts. Both the male and the female are able to turn it inside out and they do so during copulation. The male's duct leading from his sperm sacs and the female's oviduct meet end to end and sperm is transferred within seconds. Swifts, indeed, are able to complete it in flight as they glide.

For a few birds, this brevity is either unnecessary or impractical. Those that live mostly or entirely on the ground have much less difficulty in keeping their balance for they can crouch – and ostriches, storks and curassows do indeed have a long extension of the lower wall of the cloaca that acts like a penis. That of the ostrich is

Right: a male dunnock pecking
a female's cloaca before copulating, England

bright red and nearly a foot long. Ducks have a special problem. They mate on water and when the male mounts, the female may be almost submerged. Under those circumstances, sperm could well be washed away and lost unless it is deposited deep inside the female. So ducks, like ostriches, have an organ with which to inject sperm.

The transfer of sperm however does not necessarily guarantee that the male who produced it will be the father of the chicks hatched by the female who received it. It takes time for sperm to travel up the oviduct and unite with an egg. In some species, the journey may be completed in no more than half an hour. In others, sperm may remain alive in the oviduct for days, even weeks, and fertilise successive eggs as they are produced from the female's ovaries. So many males take steps to ensure, as best they can, that none of their rivals have access to a female after they have mated with her. There is an old saying in England that, in spring, you never see a single magpie. If you see one there will always be another nearby – and even a third. A male magpie, after copulation, keeps very close to his mate, accompanying her whenever she stalks across the fields looking for food, trailing after her as she goes down a hedgerow searching for a nest to raid. Many other less conspicuous birds such as chaffinches behave in much the same way. If another male approaches the pair, the male will do his best to drive him away.

The male hedge-sparrow – more properly called a dunnock, for it is not a sparrow at all – varies its marital arrangements according to the quality of the environment around its nest. If there is plenty of food, then a female may be able to rear

young with little help and a vigorous male may attract two or three females, each of whom builds her nest in his territory. If, on the other hand, the territory is poor in food, then a female may need more help in caring for her young than one male can provide. In such a case, she may be able to recruit the help of a secondary male. Her senior mate, who initially established the territory, remains her ostensible partner, singing conspicuously to defend its boundaries and mating with her frequently out in the open. The secondary male is much more retiring. The senior male tolerates having him around for he helps in rearing the chicks. But the female offers her own rewards. If she can escape the senior male's attention, she and the junior male will mate hidden quietly in the bushes out of the senior male's sight.

The senior male dunnock does his best to avoid being cuckolded by keeping a close eye on his female. But he has an additional strategy to reduce the possibility of her laying eggs to which he has not contributed. He starts his copulatory display by chasing a female and perching beside her as soon as he gets the chance. If she decides to take things further, she crouches, ruffles her body feathers and shivers her wings while the male excitedly circles around her. Then she lifts her tail and exposes her cloaca which looks like a little pink bead. The male gives it a sharp peck. Her cloaca begins to throb and may extrude a tiny white droplet which the male inspects intently. It is sperm that has resulted from a previous mating. Only when this has dropped off will he mount and give his sperm to her.

Infidelity between paired male and female reaches an even greater extreme among populations of one of Australia's most colourful birds, the superb fairy wren. The male has a shining sapphire blue head with a glossy jet-black patch on the back of his neck which extends into a bar around his eye. The female, like so many others, is a plain brown. The bush country where the wrens live not only contains little food but is poor in nest sites. Young birds tend to stay in the territories where they were hatched and to take over the nest site when one of their parents dies. So fairy wren pairs are nearly always closely related – father and daughter, brother and sister, mother and son. Incest brings the dangers of weakening the stock genetically. The fairy wrens, however, manage to reduce this by making their pair-bond highly elastic. Both male and female seek nearly all their matings outside it. A philandering male regularly visits other territories and shows off his brilliant plumage in a series of flirting displays. Sometimes he will add to his already considerable glamour by carrying a flower in his beak. He only presents such bouquets to females other than his long-term partner with whom he has established a nest. Using such wiles, he may copulate with as many as ten different mates in a season. Similarly, the female may do so with up to six different males and allow her regular partner only just enough matings to keep him feeding the family they have in their nest. The fact is, however, that none of the chicks in that nest may be his. For fairy wrens, the pair bond has become little more than a social convenience.

For one bird, remarkably, it has disappeared altogether. The aquatic warbler lives in the wetlands of eastern Europe. During the summer the birds forage in the grasslands surrounding their swamp. Here, the female aquatic warbler searches for a mate. When she finds him, they copulate but this act, far from being momentary, may last for up to half an hour. The male has huge testes and immense sperm stores and he is able to inseminate her seven or eight times during this one coupling. As a result, her reproductive system is flooded by his sperm ensuring that the egg she is about to produce will almost certainly be his. After they separate, she goes back to the nest she has constructed for herself and within the next few hours lays one egg. The next day, however, she goes in search of another male. So she continues until her clutch of up to six eggs is complete. Every one of them is almost certainly from a different father.

The production of an egg has demanded much from both parents. Both male and female have done their best to select the best partner they can find and to contribute to as many fertile eggs as possible. But much more labour will now be needed if that egg is to hatch and the young to be reared.

Above: a male superb fairy wren, Australia

8

THE DEMANDS OF THE EGG

Mammals are the not the only back-boned animals that give birth to live young. Among reptiles, certain species of snakes and lizards do. Among amphibians, there are frogs and salamanders whose babes emerge, glistening with mucus, fully equipped with four legs and capable of wriggling away as soon as they are free from their mother's vent. Among fish, species of shark produce fully formed babies, and the little guppy expels young no bigger than tea-leaves, which nonetheless can swim off to find shelter among the fronds of water plants. Only one group of back-boned animals has never evolved a live-bearing species: the birds. Every bird in the world lays eggs.

The reason is clear. Any animal that flies must keep its weight to a minimum. An egg kept within the body for days if not weeks while the embryo matures sufficiently to survive in the outside world, would constitute so heavy a load that flight would become seriously impeded. To retain a clutch of, say, half-a dozen would be an impossibility. So all female birds get rid of each of their eggs just as soon as they can.

As an ovum passes down the duct leading from a female's ovary, it is joined by a bag of yolk that will provide it with all the food necessary for it to develop into a chick. It is then fertilised by a single sperm. This, in some species, will be one of a batch delivered within the last few hours by an attentive male; in others it will come from a mass delivery of several million which arrived days or even weeks earlier, which has been stored in tubules opening from the female's oviduct, and since then has fertilised a succession of eggs.

The embryo, to develop properly, will need water and this is provided by albumen which is swathed around the yolk. Wrapped in membranes, the whole bundle moves on down the oviduct, until it reaches a section which is surrounded by a

Right: a little ringed plover with eggs, England

lime-secreting gland. This adds a shell. A little further down the duct, other glands may decorate the shell with specks of pigment, usually derived from blood or bile. If the egg twists in the duct as it moves on down, as those of guillemots do for example, then the dots and specks will smear into wriggling lines and scribbles. And then, propelled by a muscular contraction, the finished egg is ejected into the outside world

It is now very vulnerable. Its shell cannot be impregnably thick and strong for the young chick will eventually have to hammer its way out of it. The shell also has to be porous, for the chick within must be able to take in oxygen and breathe out carbon dioxide. So all eggs are relatively fragile. Since they are also packed with nourishment, they are tempting targets for hungry thieves.

The eggs of the little ringed plover are laid, unprotected, on the shingle of the sea shore. The bird has little alternative, for the shore is its home where it finds its food, but a beach has little cover and no place where eggs can be tucked away, out of sight. The plover's solution to the problem is to colour its eggs so that even when a thief is looking at them, it does not recognise them for what they are. They are brown with dots and splashes of a deeper colour which match the background so closely that they are indistinguishable from it. In shape they are pointed at one end so that the clutch of four fits neatly together and can be covered by an incubating bird, which itself has camouflaging plumage. So both the bird and its eggs effectively become invisible. The major danger facing a plover's eggs is not so much that they will be noticed as that they will be trodden on because they are *not* noticed.

Sooty terns also lay their eggs on shingle, but only do so on low sandy islands which have been thrown up by the sea so recently or so remotely that no land predators have yet managed to reach them. But terns, unlike plovers, do not nest singly. Their feeding grounds are not along a beach but in the open ocean and that is so vast, and the numbers of places where they can nest are so small, that great numbers of them must necessarily nest on the same island. On Bird Island in the Seychelles, as many as half a million pairs assemble in an immense noisy colony. They cannot remain inconspicuous as a pair of plovers can. Adult terns are not protectively coloured. They have the bright white plumage typical of many sea-birds with, in this case, a black back and a striking glossy black cap. They make themselves even more conspicuous by nesting within a foot or so of one another. That is just far enough for incubating neighbours to be out of the range of one another's stiletto-sharp beak. But why should they cluster together so closely? The colony by no means occupies the whole of the island and there is plenty of empty territory where adults could nest should they wish to be less crowded.

The terns' eggs are not totally secure. Although the island is out of reach of terrestrial egg-thieves, it is accessible to winged ones. Nesting close to neighbours, however, gives individual terns considerable protection against attack from the air. Sitting in isolation draws attention to yourself, whereas if you are surrounded by hundreds of others, you have a chance of becoming lost in a crowd. If a flying thief descends, the odds are that it will steal someone else's egg rather than yours. Not only that, neighbours will help in repelling intruders. Even so, the colony is under perpetual threat from cattle egrets. They do not alight in the middle of it where the nests are at their most dense, for there they would be under attack from all sides and be quickly be driven off by outraged parents On the margins, however, where there are still vacancies on the ground for a nest-scrape, the egrets can manage to weave a course between nests, seeking one which, for some reason has been momentarily left unguarded. Then a quick stab of the beak, an egg-shell is pierced and the nutritious yolk spills on to the shingle. Maybe, if the egret is lucky, a nearly full-term chick will be stripped of its armour and exposed to provide an even more substantial meal.

The fairy tern does not risk putting its egg on the ground. It lays only one and attempts to keep it out of harm's way by placing it in a tree. But it does not make a nest either. It balances its egg on a branch, sometimes at the point where the branch

Above: a fairy tern with its egg, Seychelles
Following pages, guillemots and kittiwakes
nesting on a cliff, Shetland Isles

divides into two, sometimes where a slight dimple in its upper surface provides minimal lodgement. The ability of a fairy tern parent to sit on its egg without knocking it off seems close to the miraculous. Others find it only too easy to dislodge them. Fodies, sparrow-like birds, swoop in if they get the chance and push off the eggs with their beaks. They fall and smash on the ground beneath, and the fodies follow and sip the yolk.

Cliffs can be just as safe from terrestrial robbers as small sandy islands. The huge precipices of the Outer Hebrides and the Orkneys, some over a thousand feet high, are colonised by millions of sea birds which settle on the narrow ledges, in itself no mean feat of aerobatics, and there deposit their eggs. Those of guillemots and razorbills are even more pear-shaped than those of plovers, with one end blunt and the other pointed. The reason for this cannot be to allow the eggs to nestle close to one another, as it is for those of plovers, because guillemots and razorbills only produce a single egg. It could therefore be a safety feature. If objects with such a shape are pushed or knocked, they will not travel far but merely roll in a circle, a valuable characteristic for eggs that may be laid on bare rock ledges. The only way to reach these ledges, apart from being lowered on ropes as men sometimes are, is by air and just as some birds fly there to lay, others do to steal. Greater black-backed gulls sweep across the face of the cliffs. If any of the nesting birds take off in fright and leave their eggs exposed, the gulls quickly take them.

One kind of cliff, however, seems to be inaccessible even from the air. How could any bird manage to reach the vertical rocks behind the curtains of a great waterfall? No hawk or heron, gull or crow can do so. Indeed, one might think that any bird attempting to fly through the tumbling tons of water would be pounded out of the air. But one species of South American swift is so small and flies at such speed that they shoot through waterfalls like arrows and reach ledges at the back where they can nest in total security.

Swifts, however, have a major problem when it comes to gathering nesting materials. Their feet are so small that they cannot land and pick up material from the ground as other birds do. Instead, they snatch feathers, fragments of dried grass and other material that floats in the air. These they fix to an out of the way surface with a sticky spittle which they produce from particularly large salivary glands. Feather by feather, layer by layer, they build this up into a small cup-shaped nest.

In Borneo, several species of swiftlets nest in caves. Those that do so in the twilight close to the entrance use a substantial proportion of feathers in their tiny constructions, but those that nest in the deepest parts of the cave, where the darkness is total, use only spittle and make nests that are, in consequence, a creamy-white. These are the edible birds' nests that are so esteemed by Chinese gourmets. One suspects that their appeal can hardly rest in their taste, for the truth is that these nests, by themselves, taste of very little and have to be cooked with something else if they are to be given a flavour.

Left: a razorbill with its egg   – 225 –   *Above: great dusky swifts nesting behind waterfall, Iguazu, Argentina*

The palm swift, a bigger relation of the swiftlets, also uses spittle in its nest. It builds on the underside of dangling palm fronds, and – in more recent times – on the vertical faces of man-made structures such as bridges. The bird begins by smearing spittle mixed with feathers on to the vertical surface of a leaf to form a pad. As this gets bigger, it is given a little ledge at the bottom. Each night, the pair roost on their growing construction, clutching the pad with their tiny hooked toes and there, precariously, they copulate. When the female is about to lay, she settles on the upper part of the pad, with her body vertical. As her egg emerges, its pointed end catches on the ledge. Once it is free of her, she holds it in position with her tail and then edges downwards, keeping the egg pressed against the pad, first with her belly and then her breast. Now she produces spittle from her beak which covers the egg. She smears it on, rocking her body from side to side, until the egg is firmly stuck to the leaf. Not content with this remarkable achievement, she usually repeats the performance and lays a second egg alongside the first.

Holes are obviously good places for keeping eggs out of harm's way and many birds have bills that are well-suited to making them. Woodpeckers chisel them out with the same action of their bill that they use when collecting wood-burrowing insects; kingfishers stab their dagger-like beaks into the mud of a river bank; parrots use their curved bills to rip away rotten wood; even bee-eaters, whose beaks are so slender they look as though they could be easily broken, will hurl themselves beak-first against a sandy cliff and do so repeatedly until they have made enough of a depression to allow them to alight and burrow from a standing position. The champion excavators of all, one might think, must surely be toucans and hornbills since they too nest in holes and have such huge beaks. But these beaks are not solid horn. Inside, they have a honeycomb structure and they are not nearly strong enough to be used for digging a hole of any kind. Instead, these birds use holes that have formed naturally or been excavated by others.

The female hornbill is very fussy about her nesting accommodation. To suit her, a tree hole has to be reasonably spacious. It must also have a chimney at the top that will serve as a bolt hole if she is attacked. Once she has selected it, she invariably improves it by plastering over any crevices or smaller holes. The material she uses varies according to her species. African hornbills use mud, the rhinoceros hornbill of Borneo uses resin, and the great Indian hornbill, the largest of all, uses mostly chewed bark, sawdust and regurgitated food. Once subsidiary entrances have been sealed, the female turns her attention to the main one. Sitting inside, she begins to plaster its sides, changing its round shape into a slit. While she is doing so, the male brings her beakfuls of food as well as more material for the plastering. Soon it is so narrow that she cannot get out. But neither can a predator of any size get in.

Her eggs, like those of nearly all birds that nest in holes, are white. Colour and decorative patterns can have no function in the darkness of a hole, and if an

Right: a female rhinoceros hornbill
at her nest hole, Malaysia

incubating bird is able to glimpse where they are, it can avoid stepping on them and breaking them. They are also much more rounded than those of birds like plovers and guillemots. A rounded egg has several advantages. It is mechanically stronger than a pointed one; it retains heat more efficiently; and it is easier to turn in order to prevent adhesions between the membranes and warm it equally on all sides.

The female hornbill will now remain, self-imprisoned, for over three months until her eggs have hatched and the young fledged. Only then will she break down the walls of the entrance, sometimes helped by her mate, and emerge.

Making a scrape in shingle or excavating a hole does not demand great dexterity. Constructing a cradle from twigs, hair, leaves, or silk requires much more delicate skills and it seems almost miraculous that a creature like a bird, with no fingers to manipulate building materials, can manage to put such constructions together. Even making the simplest platform of twigs requires considerable judgement in placing the components. Just how much, you will discover if you ever try yourself to build one that will remain stable on the branch of a tree or in the tangle of a hedgerow. Yet a bird, with one if not both feet fully employed in keeping it firmly on its perch, has to build such things using nothing more than the equivalent of a pair of tongs. A rook puts together several hundred twigs to form a deep bowl and then lines it with grass, leaves and hair; a reed warbler plaits a little cup with grass and strips torn from the leaves of reeds; eagles use the same nest site over and over again, but each year put new layers of branches and twigs on top of the old ones, so that an eyrie that has been used for a long time may be many feet high.

The hamerkop, an African bird that looks something like a small brown heron, builds one of the most substantial of all individual nests. A finished one may weigh a hundred pounds or more, measure six feet in height from its base to the top of its domed roof, and contain around eight thousand separate bits and pieces. A hamerkop when gripped by its passion to build, will use almost anything inanimate that is liftable – heavy sticks, twigs, reeds, leaves, feathers, bones, bits of fur and plastic. For a nest site, it usually selects the fork of a tree, often near a river. Both male and female work on the project, putting in a couple of hours of intensive labour, morning and afternoon. After a week, they will have built a sizeable platform. They then raise the walls, leaving a small gap on one side which will become the entrance. These are then roofed over by placing sticks upright in the walls and inclined slightly inwards, so that horizontal rafters can be interwoven between them. Great quantities of extra material are piled on top of them until the roof is as much as three feet thick and quite solid enough to bear the weight of a man. Finally, the entrance tunnel and the nest-chamber itself are plastered with mud. The whole construction is usually completed in no more than six weeks. Its huge size seems well beyond the necessities of security. It may, perhaps, serve as sign to other

Right: a hamerkop standing on top of its huge nest, southern Africa

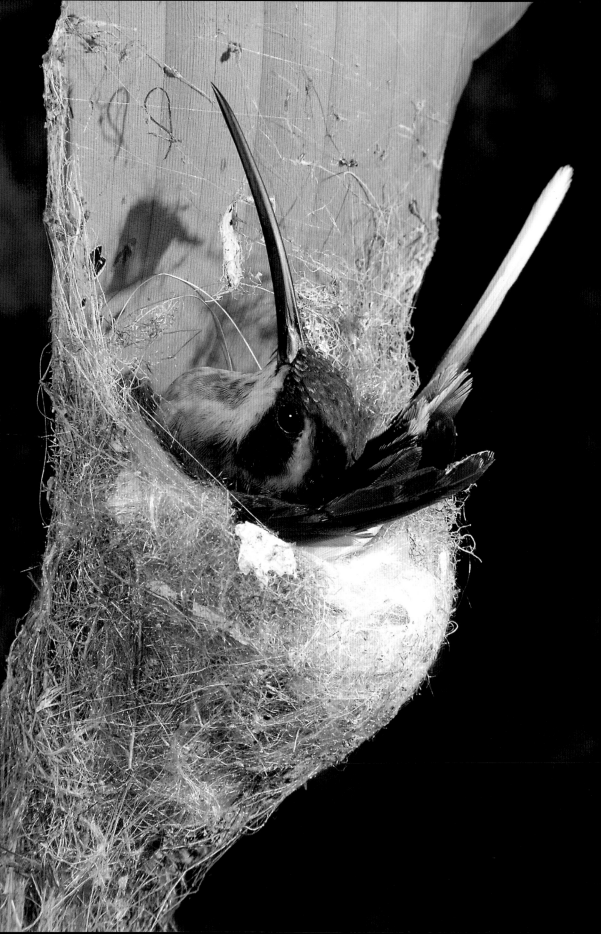

hamerkops that the surrounding territory is occupied, but if that is the case, it seems a very expensive way of sending such a simple message.

Placing sticks and other objects on top of one another in such a way that they hold together may be a subtle skill requiring considerable judgement, but it is by no means the most advanced constructional technique deployed by birds. Humming-birds collect the sticky silk of spiders' webs and use it to fashion their tiny cups, sometimes flying round and round a half-built nest carrying the end of a silken thread in their beaks so that it wraps around the outside of the nest wall. The long-tailed tit habitually decorates the outside of its domed nest with fragments of li-chen, and fills the interior with over two thousand tiny downy feathers.

Some build with mud and use their beaks, not as a pair of tongs, but as a hod and a trowel. They collect their material from the edges of pools and puddles. Sometimes they judiciously mix it with a little straw, sometimes with their own spittle. The mud-builders' constructional technique is more akin to that of a brick-layer than a potter, for they place each beakful of mud as a separate pellet on the wall they are building. As a course is completed, it is left to dry before the next course is added. Swallows and martins are expert at this, fixing their cups to walls and under eaves and lining them with just a few feathers and pieces of dried grass. Picathartes, a rare and bizarre inhabitant of the West African rain forest, with the less than at-tractive popular name of bare-headed rock fowl, builds a similar shaped cup but a foot wide and attaches it to the underside of overhanging rock-faces.

Left: a long-tailed hummingbird in its nest of spiders' silk, Venezuela

– 231 –

Above: a bare-headed rock fowl on its mud nest, Guinea

The most skilful mud-builder of all is the oven-bird of Paraguay and Argentina. It spends months each year constructing a roofed chamber very like the traditional ovens used by people in those parts. It takes considerable care to mix its building material with straw and often cow dung and the result, baked under the strong South American sun, becomes rock-hard. No beak or paw in the neighbourhood is strong enough to smash it. Nor can one reach inside, for immediately facing the entrance, there is another wall, stretching halfway across the chamber. This not only serves as a baffle against thieves, but probably also has some value as a draught excluder, for although the oven birds always build their nests so that the entrance faces away from the prevailing wind, without the internal baffles the nesting chamber would be very draughty indeed. The birds have such confidence in the security of their finished nest that they take no trouble whatsoever to conceal it. Since they usually build a new nest each year and the old ones are so durable, almost every other fence post and telegraph pole in some parts of southern South America has one of their constructions, old or new, sitting on top of it.

Above: a pair of rufous oven-birds
working on their nest, Brazil.

For the tailor bird of India, a beak is a needle. As thread, it uses spiders' silk, cotton from seeds and fibres from bark. It selects leaves that are growing and still attached to a tree, and pulls them into position so that their edges overlap. Then it pierces a hole in the margins and holding the fibre in its beak, threads it through the hole. It twists the end of the stitch into a knob which is big enough to prevent the

Above: an Indian tailor bird at its nest

thread from slipping out and does the same thing with the other end, so fixing the two leaves together. Half a dozen or so such fastenings may be necessary to convert a pair of leaves, or a single one curled round on itself, into a cup. The bird then fills this delicate chalice with strands of grass. On other continents, other warblers, the family to which the tailor bird belongs, also sew. In Australia the golden-headed cisticola, in Africa the green-winged forest warbler do so with equal skill but less fame.

The most elegant nests of all are those that are woven. To make these, the beak serves as a shuttle, threading a thin ribbon stripped from a leaf, or a tougher fibre made from a leaf's midrib, alternately over and under other strips. And this is not the only skill that is required. A bird must be able to assess how taut each strip should be pulled, and envisage the final shape the nest will take so that it can judge when the curve of the walls must be brought in or bellied outwards. In most bird species it is the female who tackles the bulk of the construction work and that is the case among the South American birds who have taken up the craft of weaving – orioles, oropendolas and caciques.

A female oropendola begins by wrapping the end of a long strip, torn from a leaf, around the dangling twig of a tree. She may hold it against the twig with one foot while manipulating the free end with her beak. She secures it by pushing the short end through the wrapping and so tying a half-hitch. Then she takes the long end and curves it round and back to the first knot. She may do this with several strips in the same place. The resulting hoop will form the entrance to the finished nest. From this she weaves the walls, hanging downwards and working from the inside, using her beak to thrust one fibre under another, and then delicately grasping the free end and pulling it tight, so creating an interlaced fabric of great neatness and uniformity. As she works, so she creates a tube that extends downwards from the entrance. When she flies away to collect another fibre, she does not take off from the open bottom of the tube, but clambers up inside it and departs from what will become the proper entrance. As the tube lengthens, she widens it so that it becomes club-shaped. Only when it is six feet long will she bring the curve inwards and seal off the bottom.

In Africa, the pre-eminent weavers are members of the sparrow family. Here – unusually – it is the male who makes the nest. The many species include the masked weavers, greater and lesser, that use their nest-weaving skills as a means of attracting a female; the grosbeak weaver that builds an outstandingly neat construction, using only thin fibres shredded from one particular grass and tucking in the edges of the side-entrance to form a selvedge; and another, the sociable weaver, that is more of a stacker than a weaver. It builds not a pouch for a single clutch of eggs, that is used for only a few weeks in the breeding season, but a giant apartment block that is occupied the year round by as many as a hundred families.

Right: a grosbeak weaver building its nest, Tanzania

Sociable weavers live in the deserts of south-western Africa. Their homes are massive stacks of coarse dry grass that may be a century or so old and weigh several tons. Some are so bulky and heavy that the branches of the tree in which they are built break under the weight and bring the whole construction crashing to the ground. The birds live in tunnels that open on the underside of the stack. Some of these lead up into nesting chambers; others are cul-de-sacs in which birds may roost. The upper surface is thatched with a layer of rather coarser stems. All the birds collaborate to build and repair this feature from which all benefit.

Such a construction could only exist in a desert where rainfall is very low. If it were soaked by heavy rains, the whole stack would rot and disintegrate. Here, however, it protects its builders from the extremes of the desert conditions. During the day, its great thickness shields those within from the heat of the ferocious sun; and at night, perhaps even more valuably, the nest chambers retain much of their daytime warmth, while outside in the desert the temperature drops to several degrees below freezing.

No matter how beautifully made, ingeniously designed or artfully suspended from the branches of tall trees, few nests are beyond the reach of all predators. The oropendolas acquire insect guardians by regularly building their nests beside those of wasps that hang from trees like papery inverted mushrooms. Such wasps have very powerful stings, certainly venomous and painful enough to deter any unprotected human being from disturbing them. Initially, when the oropendolas start building, the wasps may attack them, but within a day or so, the insects become accustomed to their new neighbours and although the birds fly busily to and fro during the nesting season and often cause quite a commotion with their squabbles, the wasps ignore them. It is only when an intruder such as an opossum or a snake makes its way down the branch towards a suspended nest, or a raiding toucan flies in that the wasps go into the attack. Only a very brave and hungry thief would be undeterred by them. Just how the insects distinguish between the birds in the colony and an intruding bird is not certain, but it is the case that oropendolas and other birds that form these associations have a strong musty body smell which others, including even closely related species, lack. The insects may themselves derive some benefit from the relationship. Caracaras, a species of hawk, regularly attack wasp colonies as do tamandua anteaters and when either appear near an oropendola colony, the birds will vigorously attack them and so protect their protectors.

The crested bellbird that lives in the Australian desert also uses insects as guardians but not so benevolently. It drapes the rim of its nest with a species of grass caterpillars the hairs of which have an unpleasant sting. To prevent them from crawling away, the bird partially paralyses each of them by giving it a sharp nip. It is true that when the chicks hatch and fledge, they eat the caterpillars and that therefore this behaviour could be regarded as assembling a larder. However

Right above: the communal dwelling of sociable weavers, Kalahari, South Africa
Right below: sociable weavers at the entrance of their nest chambers

there seems no good reason why the caterpillars should be collected such a long time before they are needed, so it does seem likely that they too are being used to scare away any burglars that get close enough to poke their noses into the nest.

Hairy caterpillars irritate and wasps sting, so both are straightforward deterrents against raiders. But some birds protect their nests by deception. Currawongs are bold and aggressive birds of the Australian bush and regular egg thieves. They destroy as many as nine out of ten of all the nests made by the blue fairy wrens. The wrens have no defensive strategy and instead, presumably, continue nesting with such determination that they produce enough survivors to maintain their numbers. The thornbill, however, a similar-sized but more soberly-coloured bird which lives

alongside the wrens, has developed a most artful defence. Having built a small dome-shaped nest, it then builds another cup on top. This is a dummy. A hunting currawong may notice it, but it will also see that it contains nothing and fly on. As a consequence, the thornbill does not suffer nearly so much as the wren from such raiders.

The Australian rainbow pitta has a different enemy and a different defensive deception. It lives in the wetter more tropical parts of the continent. In these forests, many eggs are taken by brown tree snakes. Snakes depend greatly on their sense of smell to find nests which, apparently, have a characteristic odour. The pitta, however, regularly collects smelly wallaby droppings and deposits them inside its domed nest. This is not an odd quirk of behaviour by one slightly crazed bird. As many as half the pitta population in a given area will provide their nests with perfumed protection by gathering wallaby droppings and garlanding their nests with them.

The most widespread deceptions of all, however, involve parent birds putting themselves at risk and acting as decoys by suggesting that they are a better meal than a nestful of eggs or chicks. These tricks are commonest among birds that nest on the ground and are therefore particularly vulnerable, and the most celebrated of them is that performed by plovers such as the North American killdeer or the European little ringed plover. As you approach one of their nests, whether unwittingly or knowingly, the sitting bird will leave it, often unobtrusively. When it is some distance away, it suddenly makes itself conspicuous by trailing one or even both its wings and screaming, for all the world as if it were crippled. A fox or a stoat might well be attracted away from the nest, thinking it has discovered an easy meal. Even you, if you had no notion of the tactic, might well walk towards the bird, puzzled by its extraordinary behaviour. As you approach, so it will retreat a little, drawing you with it. And when you are sufficiently far away from the nest, the crippled bird suddenly and apparently miraculously recovers and rises into the air. If the terrain is an empty and uniform one, such as an open plain or a shingly beach, then it may well be very difficult for you to return and find the precise place where the trick was first played on you, and even then you may not discover the superbly camouflaged nest.

The broken-wing trick is not the only kind of pantomime put on by ground-nesting birds. Each species will give a performance that is nicely judged to appeal to their particular predators. Sandpipers on the Arctic tundra have a version that is particularly suited to the tastes of the robber who most regularly takes their eggs, the Arctic fox. They will run from a nest with their tail held down, their wings trailing and slightly quivering, while making a squeaking noise quite unlike their normal calls. They scuttle away along a zigzag course, stopping every few yards and looking back to make sure that they have been noticed and are being followed. Their resemblance to a lemming or some other little rodent is remarkably close –

Left: a yellow-tailed thornbill feeding
its young, with a dummy nest-cup above,
Australia

and rodents, of course, are a staple part of a fox's diet. The green-tailed towhee in sage bush of the western United States makes its run with its tail lifted – and so looks more like a chipmunk, which is the favourite prey of the local coyotes. All these tricks seem at first sight to put the adult bird's life at risk, but the birds' priceless ability to rise into the air seems to save them every time.

Eggs demand not only protection but incubation. Birds keep their bodies somewhat warmer than we do, probably because they need to generate so much physical energy in order to meet the demands of flying. The embryo within the egg, on its way to becoming a young chick, does not need quite such a high temperature in order to develop but it must nonetheless be kept warm. A drop of several degrees, once development has started, may be lethal. But an embryo can neither generate nor control its own heat. It relies on its parents to do that.

The parents' task is particularly great in cold parts of the world. The snowy owl of northern Europe and the Arctic nests on the ground and may lay as many as a dozen eggs. As soon as the first appears the female does not allow it to cool as birds in gentler climates do, but starts incubation immediately, thus triggering their development. Her plumage is thick and dense, as it has to be if she herself is to remain warm in these low temperatures, but its very effectiveness as an insulator prevents her from transferring her heat to her eggs. So each year just before she lays her first egg, all the downy feathers on a patch of her belly fall out. In flight, the feathers surrounding this patch are sufficiently long to cover it, albeit thinly, so it is not normally visible, but when she starts her incubation, she lifts those feathers and tucks her entire clutch beneath them, next to her naked skin. That skin by now has become enriched with blood vessels. In such a cold part of the world, it would be dangerous to expose her eggs to the chilling air but held against her brood patch they are within a degree or so of her own body temperature. At least, that part of the egg that is next to her skin is. The side in contact with the nest may be as much as 9° C lower. She turns her eggs regularly so that all parts of it are warmed at some time, but the small thin patch of cells on the surface of the yolk that will develop into the chick always remains uppermost and nearest to her body, for the yolk revolves freely within a newly-laid egg and that side of it is slightly lighter. From now on, she will not leave the eggs until they hatch four or five weeks later. And every few hours, her mate brings food to her as she discharges the task that is so important to both of them.

Ducks and geese conserve the heat transferred to their eggs from their brood patches by lining their nests with feathers that they pluck from their own breasts. Eider ducks produce downy feathers of such fineness that human beings have for centuries stolen their nest linings to make the most luxurious of bed coverings for themselves. Gannets and boobies keep their eggs warm by standing on them with their big webbed feet. Brown boobies lay two eggs and put a foot on each and then squat, taking the weight of their bodies on the lower part of their legs which are placed flat on the nest The gannet, however, lays only one egg and stands on it with both feet – which explains perhaps why gannet eggs have particularly thick shells for their size.

Megapodes put their eggs in incubators. They are chicken-sized birds that live in tropical Australia and the islands of the western Pacific. Like chickens, they spend most of their time on the ground and find their food by kicking backwards through the leaf litter in search of seeds, fruits, and small creatures of the soil such as snails or worms. One of them, the scrub fowl, when it prepares to breed, uses the same kicking action but does so repeatedly in one small area. Several of them, working in a group, heap together a huge mound, fifteen feet high and some 35 feet across. This contains a great deal of vegetable matter. As that rots, so the temperature in the mound rises. The females that helped in building the mound, then dig burrows into it and at a depth of several feet below the surface, deposit their eggs, one to a burrow. Each burrow is filled in after laying and eventually the whole mound is abandoned. The temperature at the depth they have left the egg is usually a remarkably stable 35° to 39° C. Six to nine weeks later – the exact time depends on the precise temperature within the mound – the eggs hatch, and the little chicks, which already have huge feet like their parents, kick their way up through the soil and into the open. Within twenty four hours, they are strong enough to fly.

Megapodes, however, have the wit to take advantage of different circumstances. Other populations of scrub fowl that happen to live close to the coast do not build their mounds in the bush but go down instead to the beaches and dig burrows in the hot sand. One population on the island in New Britain has discovered that inland, there are areas where the soil is kept warm by volcanic activity and they bury their eggs there. Other members of the family, including the brush turkey and the mallee fowl, do not abandon their mounds but continue to tend them with particular care. The male mallee fowl sticks his head into the pile to assess its temperature with his tongue; the brush turkey does so with the whole of his head and neck, for that is bare of feathers. If the mound is too cool, they bring in more vegetation and pile it on top. If it is too hot, they will kick away the surface layers, removing a blanket or two, as it were.

Over-heating is as damaging to a developing egg as cooling and birds that nest out in the open in the really hot parts of the world may have to take steps to shield their eggs from the sun. If a bird is fully feathered and has no brood patch, its body warmth will not reach the egg when it sits on it. The parent bird however must itself bear the brunt of the sun's heat. One way to minimise this is for it to face the sun so that the least area of its body is exposed to the direct rays. On the Galapagos Islands, blue-footed boobies nest on bare black lava. As the sun moves round, so the boobies swivel to keep pointing directly at it. Since they are in the habit of defecating while sitting on the nest, their changes in position are neatly chronicled by straight white lines on the black lava, radiating around the nest like marks on a sundial.

Above: a brush turkey tending its mound,
northern Australia

If it gets really hot, the solution may be to stand up, keeping the egg in shadow but at the same time exposing it to any cooling wind that there might be. The Egyptian plover does that – and more. It scatters sand over its eggs, then flies off to the nearest river or pool and collects water in its beak. This it sprinkles over the sand. As the moisture evaporates, so it draws heat from the eggs beneath and cools them. The shoebill stork collects water in quantity in its huge beak and sluices its eggs with it.

All in all, keeping eggs safe from predators by weaving, moulding or digging nests, and then maintaining the eggs at the right temperature for many days, demands a great deal from a bird. And at the end of that task, it will then be faced with the even greater one of guarding and feeding the chicks. It is hardly surprising therefore, that individual birds will dodge these labours if they can do so without too much risk to the survival of their offspring.

Attempting to do this can cause rivalries – even between male and female. The penduline tit lives in most parts of eastern and southern Europe. The male attracts a female by weaving the dangling globular nest that gives the species its name. As soon as he has finished the outside walls, he starts calling, turning his head from side to side as he clings to it. Like African weaver birds, a female tours the nests made by hopeful males in the neighbourhood to assess the quality of their work. She nearly always chooses a large nest in preference to smaller ones and signals her choice by landing on it carrying a beakful of wool that will serve as a lining to it. Her preference for a big nest has a good practical justification. A large nest will hold a thicker lining. That will keep the eggs warmer for longer in her absence, so a bigger nest will allow her to spend more time away from the nest during the incubation period, gathering food for herself.

Once she has made her choice, she takes over the building work. She constructs an entrance tube that points downwards, and brings in further loads of wool or some other lining material. As soon as the nest is finished, she mates with the male who initiated the construction and begins to lay. But now the partners become competitors. If she could compel him to take over the responsibility of incubating the eggs and feeding the young, she could fly off elsewhere and start a second family. He would do the same, if he could be sure that the eggs to which he has contributed, were going to be cared for.

Six is the usual number in a clutch. As soon as the male recognises that his mate has finished laying, he will, if he can, leave her and start weaving another nest to attract another female and start a second family. Early in the breeding season, many males manage to do exactly this. But the female has her own strategy. Having laid two of her eggs, she brings in more nest lining and puts it on top of them. The male continues to mate with her and she continues to lay, concealing her eggs beneath extra layers of lining. Only when her clutch is complete does she rearrange the

lining so that all the eggs are exposed and ready for the sustained warming that will trigger their development. And with that, she leaves. The male, returning to the nest as usual, is now faced with a full clutch, so he begins the incubation. But she may never return to assist him. If he were to desert now, he would lose all his genetic investment for the season. It may be too late for him to start building a nest again. Faced with that alternative, the male will make the best of it and raise his brood single-handed. She meanwhile may have found another male hanging on his half-built nest and asking her in. So she may begin to lay a second clutch and produce twice as many young as she would have done otherwise.

The female redhead duck of North America has found a rather simpler way of increasing her output. She cannot bamboozle her mate into looking after an extra clutch, because the redhead male, like so many male ducks, is polygamous and never helps with incubation or chick-rearing anyway. Instead, before she starts laying in her own nest, she will slip on to the nest of another redhead and quickly lay an egg there. Nor is she over-fussy about who looks after her young. Canvasback ducks also often nest on the same lakes as redheads. They, however, start slightly earlier in the year and a redhead female will boldly swim up to a sitting canvasback, push her off, squat and within seconds lay an egg. The owner of the nest will not have gone far and returns very soon. But will she be able to detect that an extra alien egg has been added to her clutch? Duck eggs do not have the squiggles and blotches that enable other birds to recognise their eggs individually so there is a

chance that she won't. Sometimes she does – in which case she tips it out. But often she doesn't. In fact, some redheads lay three-quarters of all the eggs they produce in a season, in someone else's nest.

At first sight, this may seem a rather pointless arrangement, for if you have laid your egg in the nest belonging to someone else of your own species, it is very likely that someone else has laid their eggs in yours. Nonetheless, the practice may well bring real benefits. Redheads build their bulky nests on the ground close to the water's edge, often concealed among reeds but nonetheless very vulnerable. Racoons and other predators are able to raid them without much difficulty. So it is advantageous for a female to put her eggs in a number of baskets, before laying a second clutch which she will incubate herself.

Cuckoos put their eggs in baskets belonging, not to their own kind, but to birds of very different species. Europeans tend to think that the bird that appears in spring, calls loudly and unmistakably and is the very symbol of parental

Above: a European cuckoo removes a reed warbler's egg before replacing it with one of her own

– 246 –

Right: the hatchling cuckoo ejects the last remaining reed warbler's egg, Britain

irresponsibility, is the only cuckoo in the world. In fact, members of the cuckoo family are found in all the continents except Antarctica. There are a hundred and thirty-six different species of them. Forty-five of these are brood parasites, laying their eggs in the nests belonging to other species.

Some birds, initially at least, can be fairly easily tricked. The Australian grey fantail is a small insect-eating bird that builds a neat cup-shaped nest in which the female places her two small eggs. They are pale buff in colour with delicate reddish spots at the large end. The shining bronze cuckoo will fly in swiftly, while the female fantail is away feeding, and quickly drop her egg into the tiny nest. The alien egg is very much larger than the fantail's. It is also a different colour, being white with a broad purplish band at the large end and when you look inside such a nest you will immediately see the difference between it and the much smaller ones alongside it. So, you might think, would a female fantail. But she does not. It may be that cuckoos have only just started to parasitise fantails, and that the female fantails have not yet learned to look carefully at their eggs. But it is unlikely that this situation will remain this way for long. Young cuckoos become so demanding and voracious that the fantails' own offspring seldom survive. Only those fantails that begin to notice the deception and take some defensive action will manage to rear young of their own. So awareness of the danger and the need to throw out the alien egg will, over generations, become genetically ingrained. If the cuckoos are to continue parasitising the fantails, they will have to make the next move.

And that move is likely to be the disguising of their eggs. The African diederik cuckoo has done that. The same process of natural selection that led to some fantails detecting dissimilar eggs in their nests has led to cuckoos laying eggs that are an increasingly close match. Diederik cuckoos, however, like other members of the family, parasitise more than one species. They lay their eggs in the nests of red bishop birds, white-winged widow birds and several different species of weaver birds. But the bishop's eggs are blue, the widow's blotched and the weaver's speckled white. An individual female cuckoo can only lay eggs of one colour. So she will select only the species whose eggs match hers. This has led to the development of different lineages of diederik cuckoo, each specialising in parasitising one particular host.

One of their victims has, in turn, reciprocated. The masked weaver bird lives in small colonies of a dozen or so. All the females, however, do not lay eggs of exactly the same colour. Some are white, some blue, and others have a range of different coloured speckles. Diederik cuckoo females lurk around the colony waiting for chance to slip in and lay in an unattended nest. The weaver's nests are globular with just a small entrance at the side, so a female cuckoo cannot see the colour of the eggs within. Consequently, she cannot select the one that contains eggs that match the particular colour of those that she lays. The variety of eggs in the colony

Right above: a diederik cuckoo
Right below: a colony of lesser masked weavers, South Africa

is so large that the chances are that her eggs will not do so – and as a result, they will be ejected. The masked weavers are fighting back.

The lesser masked weaver is also parasitised by the diederik cuckoo. Physically, it differs only slightly from its relative the masked weaver. Both species have a black cap on their head, but that of the greater stretches farther down, over the eye. It is however, slightly smaller than the masked weaver, and indeed than the diederik cuckoo. This has allowed it to evolve a different defence.

The lesser weaves rather more elaborate nests with a downward pointing sleeve to their entrance. Their eggs are always the same colour and it is well established that one lineage of diederiks manages to match them very well. But when we set out to film a female cuckoo entering a lesser weaver's nest, we failed repeatedly. Several female cuckoos were lurking around the colony we selected, and again and again a female flew in to a weaver's nest. But each time, though she struggled hard, she failed to get inside. Later we heard that several nests had been found with a female cuckoo so firmly wedged in the entrance tube that she had been unable to extricate herself and had died. It seems that the lesser masked weavers have, within recent years, narrowed the diameter of the entrance tube to their nests and are currently winning what may be the final round in their contest with the cuckoos.

Cuckoos are not the only birds that shirk their parental duties. Cowbirds parasitise buntings, whydah birds take advantage of finches, and honeyguides do the same to white-eyes and warblers. It is easy to see why. When eggs hatch, adult birds begin the most demanding and exhausting period in their entire lives.

9

THE PROBLEMS
OF
PARENTHOOD

Sometimes, an egg speaks to another egg. That of a ground-living bird such as a quail, lying snugly in its nest beneath a bush, begins to make clicking noises as it approaches hatching. The little chick within has poked its beak into the air-filled space between the shell and the internal membrane at the blunt end of the egg and is taking its first breaths. The clicks it makes are very rapid – over a hundred per second – and they can be heard by the chicks within the dozen or so eggs lying alongside it. Those that are more advanced and on the verge of hatching will now slow down. They themselves are also clicking but much less rapidly – sixty times per second or less – and these noises have the effect of encouraging other chicks to speed up their development. As such messages circulate among the clutch, so the eggs adjust their development to one another and eventually, although the female quail laid her eggs at 24-hour intervals over a period of two weeks, they all approach the moment of emergence simultaneously. That is important for their survival.

The nest in which they lie may have become increasingly conspicuous. The visits back and forth by the female may have worn a faint path through the surrounding vegetation. The activities of hatching and the white newly-exposed insides of the vacated broken shells, even though the assiduous parent will have removed them as soon as possible, will inevitably cause a certain amount of commotion and that could have caught the attention of a sharp-eyed predator. If the hatching process were to be extended over many days, the whole clutch could be

imperilled. It is much better that they should all hatch together so that the parents can lead the whole family away as soon as possible from the tell-tale nest.

Escaping from the confines of a shell is not easy. The newly formed beak of the imprisoned chick is still soft and not strong enough to be used as a hammer. In any case, the chick does not have enough room in the egg to draw its head back and strike a blow. The best it can do is to press hard on the inside of the shell. To concentrate its effort on one particular place, its beak is armed with a tiny white spike, the egg tooth. This is sometimes placed on the end of the beak, but more usually it is on the top of the upper mandible, a little way back from the tip. The chick also has a special muscle at the back of the neck to give it added strength as it pulls its head backwards and exerts as much pressure as it can manage on the shell. In this way and sometimes by also kicking convulsively with its feet, the shell is first cracked, then broken apart so that the chick can struggle free.

Quail eggs, like those of all ground-nesting birds, contain very generous quantities of yolk and this nutriment enables the chicks to develop very fully while they are still in the shell. To give time for that to happen, the adults have incubated the eggs for a particularly long period. As a result, the infants that now stagger to their feet are far from helpless. Their eyes are open, they have downy coats which, as soon as they are dry, will keep them warm and they soon develop the ability to control their own body temperature. They do not yet have either the breast muscles to beat their wings, nor blade-like feathers that will sustain them in the air. So they are flightless. But their legs are strong enough and sufficiently well-muscled to enable them to run. Their tiny stomachs still contain a considerable quantity of yolk – in some species as much as a third of the amount that was in the egg when it was first laid – and this will sustain them for a day or so (which explains why day-old chicks of the domestic hen can survive being packed in boxes and despatched on long journeys without being fed). Quail chicks, within an hour or so of the first eggs hatching, are ready to move and their mother leads the whole brood away from the nest to look for food.

Precocious though such chicks of ground-nesting birds are, they still need protection from being chilled in the rain or over-heated in the sun, so when necessary their parents continue to care for them, gathering them together beneath their wings, close to the warmth of their body, brooding them in much the same way as they brooded the eggs. If danger threatens, a parent will give a warning call and the young will either sit totally motionless or alternatively rush to safety beneath their parents' wings.

The mother quail, like a mother chicken, helps her young to find food, scratching away with her feet and, when she finds an edible particle, putting it down in front of one of her chicks and pointing at it with her beak. Ostriches and many pheasants behave in the same way. Baby cranes and guans are marginally less able to look after themselves and have to be fed, their mothers offering morsels to them directly. Newly hatched kiwis and many shorebirds, however, are able to find their own food without any help of any kind. One way or another, all chicks of ground-nesting birds soon learn to feed themselves.

– 253 –

Following pages: great crested grebe chicks being ferried by their parents, and fed with feathers, Europe

Water birds also vacate their nests with their families as soon as they can. Jacanas pick up their newly-hatched youngsters from the floating rafts that are their nests, tuck them beneath their wings and with the tiny legs dangling untidily down their flanks, march away across the floating leaves to find a place where there are lots of insects for the young to eat. Grebes, which make similar floating nests, provide a more dignified ferry service. The chicks climb up to the back of one of their parents. Once they are on board, the adult raises its wings slightly to prevent them falling off and sails away across the lake with the youngsters peering inquisitively over the sides. The parent grebe feeds its passengers by twisting its head round and offering them morsels in its beak. But the first objects the chicks are given are not food. They are small feathers either picked up by the adult from the surface of the water or plucked from its own breast. Each little chick will swallow a great number. They may be indigestible but they are very valuable nonetheless. They accumulate in the chick's stomach. Some form a felted plug in the opening that leads to the intestine. This prevents any sharp fish bones or indigestible parts of insects that the chick will soon be eating from passing through the stomach and damaging the delicate walls of the gut. Others collect into balls with which the fish bones become entangled and held until they are largely dissolved. This practice of feather-eating will continue throughout their lives, but it must be a specially valuable precaution at this early stage.

Above: jacana chicks on lily leaves   – 256 –   *Right: a sungrebe, Guatemala*

The sungrebe, which is not a grebe at all but occupies a family entirely of its own, has the most extraordinary of any method used by birds to transport their infants. The male has pouches in his skin like saddle-bags, one on either side of his body beneath the wing. His two young hatch at an astonishingly early stage, after only ten or eleven days of incubation and they are so underdeveloped that, unlike those of any other water bird, they are blind and virtually naked. Somehow or other – no one is sure exactly how – these tiny helpless young arrive in the male's pouches, one on each side. The male is even able to take to the air with them on board. No other bird is known to carry its young with it on flights.

Duckling and goslings are even more independent than the young of chickens and quails. On the island of Greenland in the Arctic, barnacle geese nest on the face of high inland cliffs where they are safe from raids by foxes. When the young hatch, the parents fly down to the foot of the cliffs and there call to them. Down the young come, sliding and somersaulting between boulders and tumbling over sheer drops until, miraculously, they arrive at the bottom and scamper after their parents down to a stream. The departure of the torrent ducklings from their nests beside rivers in the high Andes is so dramatic it seems almost suicidal. The tiny bundles of fluff tumble out and dive straight into the rushing waters. They bob up within seconds, their little webbed feet paddling energetically, their water-repellent down making them almost unsinkable. Miraculously they manage to avoid being swept

downstream. They are so light in weight that they can ride on top of a raft of spume. They exploit the eddies, take refuge in the swirling water in the lee of a boulder and somehow or other stay on the limited stretch of river that constitutes their parents' territory.

Such independent young must be able to recognise almost immediately what is edible and what is not. They cannot, however, defend themselves so their mother must keep a close watch over them and do what she can to protect them from predators. It is therefore very important that, as far as possible, they stay close beside her and come to her when she calls. This attachment is formed early and irrevocably, with consequences that may extend throughout their lives. It may even have begun before hatching, for since the chick within the egg can assuredly hear sounds made in the outside world, it may be that even before it breaks out of the shell it has learned the distinctive sound of its mother's voice. Then, immediately after hatching, for a period which may be as brief as a few hours but in no case is longer than a few days, the ducklings and goslings learn to recognise the appearance of their parents and the sound of their voice. Those memories will stay with them for the rest of their lives and greatly influence in due course their choice of mate.

The link is tested immediately. Their parents, having been guarding and incubating the eggs, are now hungry and anxious to get to their feeding grounds. So they set off on what may be a long journey. Since the young cannot yet fly, it has

Above: torrent ducks and young, Chile

necessarily to be on foot. The babies trail after their parents, led by the sound of their voice and the vision of their legs. Shelduck families, which nest in disused rabbit burrows or hollow trees, waddle determinedly down to the nearest stretch of water that can provide them with food. There they assemble in groups. As more and more families emerge, a large flock forms. The young are vulnerable to gulls which swoop down and snatch one if they can. The adult shelducks on guard react ferociously and drive off the gulls with vigour. Indeed, they are so competent at doing so that one or two adults can effectively protect a whole raft of ducklings. So as more adults arrive, others who may have been guarding the creche for a day or so, leave to feed in more distant waters.

Young shelduck, with such protection, can fend for themselves. The young of terns, flamingos and king penguins, all of which are also left in creches, cannot. So the parents of these birds have to return regularly with food and it is then that the imprinted sound of the parental voice becomes essential to the chick. In the incessant din of a king penguin colony, it seems impossible to a human ear that the voice of any one bird could be disentangled let alone recognised individually. Yet it has been shown incontrovertibly by marking chicks and adults, that both parent and offspring have no difficulty in recognising one another. This does not mean, of course, that a young penguin will not try its luck. An adult returning to the creche with a full crop will be importuned by a crowd of youngsters, desperate for food, shrieking and prodding with their beaks to try to persuade the adult to disgorge a

Above: an adult shelduck with a crèche
of ducklings, England

fish or two. But the adult knows its own young and nearly always manages to deliver the food where it should belong.

Ostrich chicks are also herded into groups by adults, but for a very different reason. The adults seem positively eager to take on the responsibility of caring for them. If two pairs meet, each with its own brood of a dozen or so, there is often a dispute. One pair drives the other away and the victors chivvy the losers' chicks into their own flock. This may happen several time in the season so that a particularly vigorous and assertive pair may end up with a troop of youngsters several hundred strong. The different families of chicks are likely to be of different ages, so in spite of the fact that the pair's own clutch hatched almost simultaneously, the group they now shepherd about will contain individuals of very varied ages and sizes. This behaviour may well be to the advantage of the true chicks of the guarding pair, for they are likely to be safer surrounded by other youngsters than they would be on their own.

Of all ground-nesting birds, the most swiftly independent is the infant megapode. Its incubation period is particularly long – sixty to eighty days. The egg, being buried in a mound, does not have to withstand being rolled around and sat on, as eggs in nests must be able to do. Accordingly, it has one of the thinnest of shells and the chick is able to break its way out without much effort. Indeed, although some three weeks before it hatches, a little egg tooth begins to develop on its beak, this comes to nothing and soon disappears unused. The megapode chick is the only bird that manages to break out of its shell without the assistance of such a tool. But the task that it next faces is a very exhausting one. Above it lies a foot or so of earth. Lying on its back, it kicks out with its feet which, like those of its parents, are disproportionately large. As it loosens the earth, it humps its back and wriggles so that dislodged soil is pushed beneath it and it slowly moves upwards. Short bouts of frenzied activity are interspersed with long periods of rest during which it recovers its strength. All this activity is fuelled entirely by the remains of the huge yolk which lies within its infant stomach. It will take several days to dig its way up to the surface, but by the time it does arrive there, it is a totally independent individual with a full complement of feathers that are so well developed that it is able to fly immediately.

Birds that nest in trees or holes do not need to vacate their nests as speedily as those that lay their eggs on the ground for their nurseries are much more secure. Accordingly, they can rear their young in a different way. Their chicks hatch at a much earlier stage in their development. They are naked and their eyes are closed. Their legs are only feebly developed and they are unable to stand. In fact they hardly look like birds at all, more like some kind of grub. Their best-developed organs are their gut, gizzard and liver, for their priority at this stage is simply to take in and process food. Hatching early made it unnecessary for the female to provision

her eggs with large yolks and indeed, such chicks as these, unlike those that left the nest early, have by this stage very little yolk left in their stomachs But now their parents can deliver food directly to them and they grow much more quickly than independent ground-living chicks manage to do.

Some adults provide their offspring with a special pre-digested food. The pelican produces a kind of fish soup, and shearwater parents turn their meals of plankton and small fish into a rich oil which they regurgitate for their young. Pigeons produce a very special secretion from their crop which is rich in protein and fat and is known as 'pigeon's milk'. Unlike mammalian milk, it is produced by both male and female. Most birds, however, provide the same kind of food for their young, offered in suitably small portions, as they themselves eat, though among those that have a mixed diet, it will be biased towards proteins which are so important for physical growth.

The chicks beg for their food. Young herring gulls, sitting in the scrape on the ground that passes for a nest, point their mouths at a red spot on their parents' beak. Even thrushes so young that their eyes are not yet open will suddenly crane their heads upwards and open their beaks at the slightest vibration that might suggest that one of their parents has arrived with food. Their gapes, at this stage in their lives, are a bright yellow with swollen flanges and doubtless indicate to the parents where the beakfuls of food should be delivered. The flanges are extremely sensitive, so that if for some reason a chick has ceased to gape the lightest touch of its beak flange will stimulate it to rear up and open its beak. Such a signal is more

necessary for those parents whose chicks sit deep in the recesses of a hole. Gouldian finches nest in this way. They are the most colourful of their family and so presumably they are particularly sensitive to colour. Their chicks certainly exploit it. They have large knobs on each side in the corners of their mouths which are an opalescent green and blue and reflect light filtering down into the depths of their nest hole in such an effective way that they look almost luminous.

Colourful gapes can tell parents more than just the location of their chicks. They can also indicate which chicks among their brood have just been fed and which are still in need of a meal. The gapes of young linnets are red, due to the blood in the vessels just beneath the skin of the throat, but when the chicks are given a meal, much of that blood is diverted to their stomach to collect the digested nourishment. The gapes of those chicks that are still hungry are thus the reddest and it has been shown experimentally that the parents use this difference in colour to determine which of their brood will receive the next delivery of food.

Left: white pelican chick reaches inside its parent's bill

Above: a male linnet feeding his chicks, Europe

The chicks of birds that, like the cuckoo, dump their eggs in the nests of other species have to take account of this signalling system if they are not to starve. A newly hatched African whydah whose mother sneaked it into the nest of a waxbill seems an obvious interloper for whereas the young waxbills are naked and pink, it is covered with a mauve down. But the waxbill parents never see it like that. As soon as one of them alights at the nest, the young reach up with open mouths – and then they all appear to be virtually identical, for the young whydah has developed mouth spots that closely mimic those of the young waxbills. The screaming cow-bird in Argentina enhances its mimicry even more profitably. It leaves its eggs in the nests of several other species, including those of other non-parasitic cowbirds. All its victims, however, have chicks with plain red gapes. But the gape of young screaming cowbirds is a far more intense red and permanently so. In consequence they are always fed before their foster parents' own brood.

Linnets do not start incubating their eggs until their clutch is complete and as a result all hatch within a relatively short time. Their parents then rely on the colour of the chicks' gapes to ensure that the youngsters are fed in turn. Rosella parrots in Australia, however, have a more difficult problem. They start incubation immedi-ately the first of their clutch of half a dozen or so is laid. Their eggs therefore hatch over a period of at least eight days and to begin with there is a great difference in size between the chicks, with the oldest sometimes weighing five times as much as the youngest. But rosella parents distribute food to their offspring so equally that by the time the chicks are ready to fly, they are all the same size.

The young of European coots are also treated even-handedly. If one of them be-comes too greedy, squawking loudly and pushing the others out of the way, its par-ent will pick it up and chastise it by shaking it vigorously. It may even be given a thorough ducking in the water and become so distressed that it takes refuge in the reeds while its parent feeds its siblings. It may never return. This happens with such regularity that most coot broods are quickly reduced from seven to three.

Coots in America, which belong to a different species, behave somewhat differ-ently. Their chicks have a brilliant orange tuft of down on their heads beside a bald patch which is bright scarlet. When they beg for food they flaunt this brilliant head gear, nodding frantically in front of their parents. The brightest are fed first. These are likely to be the strongest and most vigorous of the brood and therefore the ones most likely to survive the trials that lie ahead. So it is better that they should get as much food as they need rather than to spread what food there is more thinly and perhaps inadequately among them all. The result is that American coot families also reduce quickly in size. Almost a third of the chicks that hatch each year die from starvation.

Such a strategy of differential feeding is practised by several different groups of birds. All boobies adopt it to varying degrees. The blue-footed booby lays two or

Right above: a coot punishing its chick, Europe
Right below: an American coot feeding its chick

three eggs which hatch about four days apart. There is consequently a considerable difference in size between the chicks. In years when there are plenty of fish around, the parents may be able to feed all three, but if fish are hard to find, the youngest chick will not be able to compete with its elders. As they grow stronger, it gets weaker. Eventually it will be pushed out of the nest and die. The system seems unnecessarily harsh, but the boobies at the time they are laying eggs, have no means of predicting how abundant fish will be in the weeks ahead. If they lay only a single egg and there proves to be an abundance of fish, they will have missed a precious breeding opportunity. A second or third egg is a small stake in a gamble that could double or even triple the number of offspring a pair raises in a season.

Most birds of prey also lay more eggs than they can raise, feed the eldest preferentially and allow it to harry its younger sibling so unrelentingly that it dies. The winner will then usually eat the loser, so the nutriment invested by the parents in the extra egg and the food they gave to the nestling it produced, is not wasted. The macaroni penguin has a strange variant of this practice. It also lays two eggs, but the first is smaller than the second, hatches later and seldom survives. Why there should be this reversal in the fate of the first-laid egg is uncertain. It may be that macaroni eggs and chicks are at the greatest risk early in the breeding season when the colony is just establishing itself. At that time, adults are quarrelling viciously among themselves and skuas lurking around the outskirts of the colony have the

Above: a blue-footed booby feeding its bigger chick while the smaller starves, Galapagos

– 266 –

Right: a harrier hawk steals a chick from a weaver bird's nest

best chance of grabbing an egg or a newly hatched chick. Indeed about half the first eggs laid by the macaronis are lost. So in the unsentimental practice of natural selection, it is better for these gamblers to place a heavier bet on the second runner.

The chicks of other birds, even though they hatch more or less simultaneously, will nonetheless fight among themselves for survival. Young Asian blue-throated bee-eaters even have a special weapon for doing so. To begin with, there may be seventeen of them in a clutch. Each has a sharp downward-pointing hook on the tip of its bill with which it slashes at its still-naked siblings in the darkness of the nest-hole. Two or three are likely to be so badly wounded that they die. The mayhem will continue for about two weeks, by which time those still in contention have started to grow feathers. These protect them from further injury. The hooks on their beaks by now have either worn away or dropped off. The survivors – and there are very seldom more than three – then settle down amicably together.

The dangers facing most chicks, however, come not from their siblings but from predators. In Europe, magpies regularly raid the nests of blue tits and other garden birds. Toucans, in South America, will eagerly vary their diet of fruit with a small defenceless bird if they get the chance. In Africa, the harrier hawk or gymnogene has such a taste for chicks that it has evolved special adaptations to obtain them. Its legs are unusually long and double-jointed. It is thus able to reach into a nest-hole, or up into a suspended weaver bird nest, even one with a long protective entrance tube, and drag out the struggling chick.

Many snakes are very agile climbers, using the scales on their underside to catch on the bark, winding themselves through the twigs to slide into a nest and grab a chick. On the ground, rats and stoats, cats and squirrels, foxes and racoons, all find small chicks excellent meals, packed with nourishment. Concealment of a nest, wherever possible, is therefore of great importance to birds.

Parents may take steps to avoid leaving clues to their nest's position. Many chicks, in their first days, extrude their faeces enclosed in gelatinous sacs. The adults often swallow these and some seem almost to relish doing so. Perhaps they contain some residual nourishment that an adult's more robust digestion is able to extract. At any rate, a parent bird, having fed a chick, will often prod its baby's

Above: a racer snake taking a warbler chick, California

Right: a parent blue tit removing a faecal sac from its young, Britain

rear, encouraging it to raise its rump and defecate, and when it does so, its parent collects the little white sac as it emerges. As the chick grows older and its digestion becomes more efficient, its faeces are less likely to be eaten. Instead they are carried some distance away and dumped. The lyrebird takes no chances. If there is a stream within reasonable distance, it will drop its load in the water. If there is not, it buries it. A potoo chick is just as well-concealed as its parents but it would soon be detected if it whitewashed its perch. It is, however, possessed of a phenomenally powerful anal squirt. Its faeces are not so much droppings as missiles which land several yards away from where it sits. Birds that nest in holes are also usually faecal squirters. Kingfishers and hornbills will back up to the entrance of their holes, lift their tails and eject a stream with sufficient power to ensure that, as far as possible, the tell-tale streak of white falls clear of the tree trunk or river bank. This removal of

faeces from the nest is not only protective, it also keeps down the risk of disease and infection by parasites.

As chicks grow, they make greater and greater demands on their parents. In the tropics, young cranes and storks, sitting on their stick platforms out in the open beneath the blazing sun, are in real danger of getting seriously over-heated, so some parents bring stomach-fulls of water and regurgitate it over the young as a cooling shower. Herons will provide the same relief but in a less sanitary way and defecate over their babies. But the greatest, the most insistent of the demands of the young is for food.

The appetite of chicks can be gargantuan and seemingly insatiable. A great tit, which feeds its young with beakfuls of insects, may deliver food to its nest nine hundred times a day. The foster-parents of a young European cuckoo, driven into a frenzy of food-collecting by the super-stimulus of the chick's enormous yellow gape, thrust food into it with such assiduity that in three weeks the chick increases its weight fifty times. Nevertheless, if both parents share the task, the labour of feeding and protecting a young family is not overwhelming. Where conditions are at their most favourable, the whole job can even be done by a single adult. That usually proves to be the female. The males then pay no attention whatever to such offspring they have fathered and devote their lives largely to display, as do birds of paradise, pheasants, manakins and other polygamous birds.

Above: a white stork cooling its nestlings with a shower of regurgitated water, northern Europe

– 270 –

Right: a shoebill stork shading its young, Zambia

But not all birds live surrounded by such abundance. Some bird live in territories that are so poor in food that they can only succeed in rearing a family if they get help. And that often comes from previous generations of young.

Such assistance is not a demonstration of unselfish altruism on the part of the young birds. There may be no room in the surrounding area for new territories. If a youngster tries to establish one, other residents will fight to repel it. Better to stay where it is welcome, as it most certainly is when its parents are desperately searching for food for their new family. Furthermore, by doing this, it is at least aiding the survival of those genes that it shares with its younger siblings and that, while not as good as parenting its own brood, is better than not breeding at all. Eventually one of the old birds will die, and then one of the helpers will be well placed to mate with the survivor and inherit the family property. This habit, which was recognised only recently, is now known to be widespread.

Jays in the arid impoverished scrub-lands of Florida behave in this way. So do babblers in the deserts of Arabia and white-fronted bee-eaters in Africa where conditions are so variable from year to year that food can unexpectedly become desperately difficult to find. And so do almost half the bird species that live in the

eucalypt woodland of south-eastern Australia also behave in this way. At first sight, this seems odd, for the bush here does not appear to be particularly impoverished. The cause, however, may lie in the equability of the climate. Here there is no sudden spring to bring a surge of new leaves and a flush of insects to feed on them, as there is in the woodlands of the northern hemisphere. There is therefore no sudden superabundance of food at any time. The birds in residence have no difficulty in feeding themselves, but when they begin to breed and suddenly have four or five additional mouths to supply, finding sufficient food becomes very difficult indeed. Kookaburras live in these woodlands, subsisting on snakes, lizards and small rodents. Their fledged young not only stay with their parents to help with gathering food for the new broods, but also assist them in defending their feeding territory. They do so, primarily, by sound – that hysterical laugh for which the kookaburra is famous. The louder the laugh, the better guarded the territory, so young birds join their parents in laughing choruses.

White-winged choughs also live in these Australian woodlands. They do not become sexually mature until they are four years old and until that time many of the immature birds stay with their parents. Their staple diet is grubs and finding them is a very labour-intensive business. They have to dig holes in the ground that may need to be as much as eight inches deep. Older birds are very skilled at this and can

Above: a family of Arabian babblers huddle and preen together, Israel

Right above: a white-winged chough at its nest, and (below) feeding its newly fledged young, Australia

quickly assess which places are likely to yield grubs. Young birds take some time to learn this and do not therefore hunt with such success. As a consequence, they continue to rely on their parents for food until they are about eight months old.

With food so scarce and hard to collect, the breeding season inevitably precipitates a crisis for the choughs. A pair usually lays four eggs. Their only chance of raising any young at all is with help. With two assistants, they may be able to gather enough food to rear a single chick. To raise all four, they will need at least eight helpers. Such a powerful group may then be able to expand its territory. A family gang – a pair with up to a dozen of their immature young – will raid the territory of a neighbour and, if they can, destroy their nest, tipping out the eggs or young and pecking to pieces the beautifully crafted mud bowl. And the warfare does not end there. If at a later stage when the nestlings have fledged, two families encounter one another, there will be great displays of aggression. But while some adults are squabbling, others may be engaged in kidnapping. They approach a newly-fledged nestling of their rivals' group and entice it away with offers of food. If a youngster is lured across, it will within half an hour or so accept its erstwhile rivals as its own family. It may be that at this early stage in its life, it has not properly learned the identity of its own family, for a month later such kidnappings never happen. But for the family which has recruited an extra member, there will be more helpers to rear the next generation.

As nestlings approach the end of their infancy, they begin to fledge. Proper feathers begin to appear, displacing any down that the chick may have had initially. Each develops from a small pocket in the skin. At first it is a soft rod encased in a sheath. It grows very rapidly indeed and soon bursts free and expands. The core of blood vessels and living cells is then absorbed back into the skin and the quill, now hollow, together with the barbs on either side, hardens. It is now as dead as a toe-nail.

Most youngsters at this stage have still not achieved their adult weight and will not do so until several weeks after they have left the nest. About a fifth of all species, however, are heavier than their parents. A newly-fledged white pelican chick has in fact reached full size but it weighs nearly forty per cent more than an adult. This is because growing tissues contain more water than mature ones. A few – among them albatross, parrots, owls and kingfishers – owe their extra weight to an additional cause. They have built up large food reserves, thanks to the lavish feeding of specially attentive parents. Those of the dusky shearwater of New Zealand weigh twice as much as an adult and it is these generous fat reserves beneath their skin that cause them to be called mutton birds.

The young bird, in its new but untried coat of flight feathers, sitting on a ledge, peering out of its nest hole, or squatting precariously on a nest high in a tree, now faces its first real move to independence. Many need little encouragement to leave the nest which by now has become seriously overcrowded. Those such as thrushes,

Right: a ruby-throated hummingbird nestling exercises its wings, perched on the rim of its nest, North America

blackbirds and robins that, as adults, will collect their food from the soil or the leaf litter, flutter clumsily to the ground or perch unsteadily in bushes, pleading with their parents to continue supplying them with food. They are still not skilled at getting into the air and are easily caught by ground predators such as cats. Many will certainly perish.

Those that sit in nests high in trees or on the ledges of cliffs, if they are to avoid a catastrophic crash, have to be reasonably competent in the air from the very moment that they launch themselves into it. Young eagles spend hours every day beating their wings and bouncing up and down on their nest, strengthening their wing muscles and doubtless generally getting the feel of what it is like to be airborne. Young hummingbirds also practice but, to begin with, do so more cautiously, holding on to the rim of their tiny nest-cup with their feet to prevent themselves being swept upwards by their whirring wings before they are quite ready for it.

Some youngsters are so hesitant that they have to be persuaded to take to the air. The parent shearwaters, having fed their young in their nest burrows with such generosity for sixty days, simply stop doing so. After six days without eating anything, the chicks, now living on their fat reserves, come to the burrow mouth for short periods and exercise their wings, something that they cannot do in the restricted space of the nest chamber. It is only after twelve days of starvation that they feel confident enough – or hungry enough – to spread their wings and launch themselves into the air.

Peregrine parents use the carrot rather than the stick, the promise of food rather than the threat of starvation. It is usually the female who takes this responsibility. Instead of bringing newly-caught prey back to the nest, she settles in a tree nearby and, with her catch still in her talons, calls to her three or four young. Having attracted their attention, she flies from tree to tree carrying the meal that they seek, calling as she does so. And then, at last, one of the young daringly launches itself into the air. Flapping rapidly and clumsily, the young bird lumbers through the air towards its meal and eventually claims it.

But this is only the first lesson. Each chick now takes up its own perch where it will be fed. As they grow in strength and confidence in the air, the female makes the collection of food a little more difficult. She holds on to the prey and persuades the young to try to take it in mid-air. When the youngster comes towards her, she rolls over on her back so that it can take the prey from her talons. The next lesson, which requires still greater aeronautical skills, is how to catch things in mid-air. She flies ahead of one of her young and drops the food so that the youngster has the chance to swoop and grab it. If it misses, she or perhaps her mate who may be keeping a watchful eye on these tutorials and is flying below, will catch it before it hits the ground and then they give the youngster a second chance.

So at last, young birds of all kinds achieve their independence. The labours of their parents in selecting the best possible mate, placing the eggs in as safe a position as can be found, and working unceasingly for days on end to provide food for extra mouths has, at last, been rewarded. A new generation is in the air.

10

THE LIMITS OF ENDURANCE

No other group of back-boned animals has colonised the earth as extensively as the birds have done. Amphibians, with their moist skins, must remain within reach of water. Reptiles have escaped from that restriction and most have become entirely terrestrial, but lacking the ability to generate their own body heat internally, they cannot survive in regions that are permanently cold. Mammals, with their warm blood and insulating fur, have succeeded in making their homes amid snow and ice and a few have even taken to the seas. But they too have their limitations. They cannot, unaided, colonise remote oceanic islands. Such specks of land, fertile and rich in food though they might be, were even beyond the reach of human beings until the invention of sea-going craft a few millennia ago. Birds, however, have long since broken all these barriers and reached all these places. They can tolerate the most extreme cold for their body temperature is permanently high, excellently controlled and efficiently conserved by their plumage. Some are superlative swimmers and divers, capable of descending to depths of a thousand feet, and with their mastery of the air, they can reach the most distant island and ascend towards the very frontier of the earth's atmosphere.

But there is one element that neither birds nor any other animal can do without. Liquid water. Deprived of that, their bodies can neither digest their food, circulate the nourishment they derive from it, nor get rid of the poisonous waste it generates. Some, such as eaters of fish or fruit, manage to extract all the water they need from their diet, but in dry environments the food birds find is inevitably poor in liquid. For them, water is an imperative and its absence a major limitation.

The deserts of Africa and the Middle East are over great areas so dry that they have no permanent vegetation whatsoever. There is nothing but scorched rock, pavements of wind-polished pebbles and huge dunes of sand. Rain does occasionally fall, but it is very infrequent and sporadic. When it arrives, however, the desert, almost miraculously, springs to life. Seeds that have lain dormant in the sand maybe for decades suddenly germinate. The tiny shoots grow rapidly. They flower within a week or so, but with no more rain to sustain them, they wither and die. By then, however, they have released thousands of seeds which in many species are as fine as dust. And it is these that constitute the main food of sandgrouse.

Sandgrouse are not grouse at all, but distant relatives of the pigeons which they resemble both in size and general shape. The seeds they live on are so tiny that each bird has to collect some eight thousand a day if it is to be reasonably well fed and even that huge number does not weigh, collectively, more than about a gram and a half. They collect them with rapid stabs of their beak. Working like manic sewing machines, they move over the sands, picking up several seeds a second. The seeds, however, are very dry and cannot supply the sandgrouse with the moisture they must have, so every two or three days, the birds must stop eating and find a drink. If the weather is particularly hot, they may need to do that once a day. Waterholes are few, but the birds will fly as far as 50 miles to visit one.

The patchy nature of rains in the desert and therefore of its seeds means that in many deserts sandgrouse can have no settled homes or territory. They are permanent nomads, wandering across the sands and nesting, when the time comes,

wherever they may happen to be. They lay their eggs, usually in a clutch of three, in a shallow scrape in the sand. As soon as they hatch, the chicks leave the nest and start collecting seeds for themselves. But they too must drink – and they cannot fly. Water must be brought to them and it is the male that takes on the task.

Parent birds of some species ferry water to their young by carrying it in their crops, but the male sandgrouse needs all the water he can accommodate in his crop for his own survival on such long journeys. He does, however, have another and unique way of carrying it. The feathers of his breast and underside are covered on their inner surface with a mat of fine filaments. When he arrives at a waterhole, he rubs his underside in the sand and dust, so removing any water-repellent preen oil there may be on those feathers. Then he moves to the water's edge. First he slakes his own thirst, sucking in water and lifting his head to gulp it down. Then he wades into the water, lifting his wings and tail to keep them clear of it, and begins to rock his body to and fro so that his belly feathers get thoroughly soaked. The mats of filaments on them absorb water like a sponge. He may stay in the water, no doubt luxuriating in its delicious coolness, for several minutes, but usually the huge numbers of other sandgrouse flying in from all over the desert for their daily drink create such a press at the water's edge that he cannot hold his place for long and he takes off.

His liquid cargo, held between his feathers and his body, is well protected against evaporation, but even so most of it will have gone if he has to fly much more than 20 miles. When at last he lands near his chicks, which are scattered over the sands searching for seeds, they run to him. He lifts his body high and they extract the liquid from his belly feathers like mammalian babies suckling milk. His load delivered for the day, he walks away and dries himself by once again rubbing his belly in the sand. He performs this service every day for at least the next two months until his chicks have completed their first moult and can fly and collect water for themselves.

Finding water might not seem to be a problem for those birds that live in the Rift Valley of eastern Africa since lakes lie along its length. But these lakes are not like others. The streams that in the rainy season drain down the sides of the Rift and flow into the lakes dissolve salts from the volcanic ash and lava over which they pass. As the lake water heats in the baking sun, so it evaporates and the salt content rises, with the result that most of these lakes are now much more briny than the sea. But one of them is so heavily saturated that around its margins the salt is solid. This is Lake Magadi. It is different because it acquires its salt not only from streams but from underground. The titanic forces, deep within the earth, that are pulling Africa apart, creating the Rift, have also caused lines of faults along its margins and across its floor. Several volcanos rise from them as do hot springs which bubble up through the floor of the Rift. One of them feeds Lake Magadi. As the superheated

water is forced up through the rocks, it dissolves sulphates and carbonates. When it reaches the surface and cools, these salts solidify and form white curds that glint and shimmer in the sun. It must seem like some cruel joke to a weary traveller approaching the lake on foot and tortured by thirst, that the water ahead should appear to be covered by ice. When he reaches it, he finds that the curds and plates covering the mud around its edges are so caustic that they will burn the skin from his flesh and if he tries to drink the tepid stinking liquid beyond, he will retch.

Flamingos feed on all of the salt lakes of the Rift at various times, and share the food in and around them with several other species of birds, but few except flamingos can tolerate the conditions at Magadi. Fish cannot survive in its waters except near some springs at the southern end where the newly erupted water is very hot but slightly less salty.

Algae and brine shrimp, however, can do so, and because so few other organisms can survive in these conditions and compete with them, they proliferate in vast quantities. And they, in turn, are food for flamingos. Two species are here, the greater flamingo, which stands nearly five feet high, and the lesser, which is half the size and constitutes the overwhelming majority. Both have webbed feet which spread their weight and enable them to walk over mud without sinking too deeply into it. Their legs and feet are scaly and not affected by the caustic salts. The greater flamingo collects shrimps and worms by plunging its head and often most of its neck deep in the water and walking forward, ploughing through the mud. But it is the lesser flamingo that has become the most extremely specialised for life in this forbidding environment.

It lives predominantly on the blue-green algae. These microscopic plants float in the upper surface of the water and the lesser flamingo collects them with one of the most complex beaks possessed by any bird. When feeding, it lowers its long neck and holds this beak, upside down and pointing backwards, just beneath the surface of the water where the algae it seeks congregate. The lower mandible is bulbous and has a honeycomb of air-filled spaces which cause it to float, so minimising the muscular effort needed to hold the beak in the right place in the water. The bend in the middle of the beak, that gives the flamingo its equivalent of a Roman nose, is of particular importance. Were the beak to be straight, then when it opens, the gap between the mandibles would steadily increase from the corner of the gape to its tip. With a bend in the middle, the flamingo is able to separate its mandibles just slightly so that the distance between them is almost the same along nearly all of its length. There is therefore no danger of a feeding bird taking in larger objects than it wants. Internally, the edges of the mandibles are lined by hinged horny plates. The bird's tongue acts as a pump, moving extremely rapidly backwards and forwards. As the tongue retracts, the hair-covered plates are swept down flat and water is drawn in. As the tongue pushes forward again, the plates lift, water is expelled and

the algae and shrimp are strained off by the hairs. Backward pointing spines on the palate and the tongue guide the food particles into the throat. The bird is thus able to swallow its food with the minimum of salty water. Pumping in mouthfuls twenty times a second, a lesser flamingo can filter twenty litres of water a day and extract from it sixty grams of food.

As the dry season approaches, the heat intensifies. The water level of the lake begins to drop and the birds' feeding grounds shrink. Worse, land predators such as jackals, hyenas or even lions, can now reach them. The flock's behaviour changes. They begin the preliminaries of courtship, forming platoons and marching past one another with parade-ground precision, flicking their heads from side to side. They salute one another by flashing open their wings to reveal their red coloration which is there at its most intense. And then suddenly, overnight, the whole flock disappears. This does not happen every year. Nor is it possible to predict with accuracy exactly when in any year it will do so. But the birds have gone away to breed. The same thing will be happening simultaneously at all the other salt lakes in the Rift.

Above: lesser and greater flamingos
on their mud nests, Kenya

Where the flamingos go was, for many years, a mystery. Only in 1954 was it discovered that they flew to the most savage and, to human eyes, the most forbidding lake of all, Lake Natron. Like Magadi, it is fed with salt from subterranean sources which solidifies around its margins, but it is many times bigger, being ten miles wide and forty miles long. The algae it contains flourish in such quantities that they stain the soda pink. Now, in the dry season, it gets so hot that surface temperatures reach 65° C. It is so big that the mud flats in its centre are, at this time, beyond the reach of land predators.

Three million pairs of flamingos assemble in an immense flock in a shallow area so far from the shore that they are almost out of sight. Here they build mounds of mud with a shallow depression on top in which they lay their single egg. The nests have to be sufficiently tall to keep the egg above any spray that the wind might blow up from the surface of the water. If that caked the egg with salt, the chick within would suffocate. The parents feed the chicks as pigeons do, with a milky secretion which they produce from their throats and drip into the chick's uplifted bill. As the dry season advances and the water level continues to fall, it becomes possible for a jackal or some other predator to splash through the shallows and reach the nests. By now, however, the chicks are strong enough to walk. Soon they begin to leave the nesting flats, marching in long columns across the salt to look for deeper water where they might find food for themselves. The journey may be a long one.

Above: an African fish eagle seizes a young lesser flamingo, Lake Nakuru, Kenya

Right: a macaroni penguin colony, Crozet Islands, southern Indian Ocean

In years when the rainfall is low, the shallows through which they trudge are only a few inches deep and the water has become so saturated with salt that it solidifies and forms hard rings around their legs. The weight of these anklets becomes so great that eventually, many chicks become exhausted and collapse and die on the salt. In such years, whole generations of chicks are lost. That is one of the penalties of colonising one of the hottest places on earth.

The coldest places on earth lie in Antarctica. Penguins standing on ice floes have become the very symbol of frigidity. In fact, eleven of the seventeen species in the family live in gentler climates where they seldom if ever see ice – in Australia, South Africa, the shores of Peru and even on the equator in the Galapagos. Some, however, endure very bleak conditions indeed. The island of South Georgia lies just outside the Antarctic Circle, so it does not totally lose sight of the sun even in midwinter. Nonetheless its jagged mountains are cloaked in great glaciers which sweep down into the sea and raging gales raise huge seas which crash into its coast. In winter, snow drifts six feet deep cover its slopes and the temperature falls to 15 degrees below zero. Macaroni penguins arrive on this inhospitable island in mid-October, the southern spring, and assemble in vast colonies two hundred thousand strong. The males come first and claim a territory. The females follow a few days later, and after twelve days or so on land, begin to lay. Male and female take turns incubating the eggs in ten-day shifts and after a little more than a month, the eggs

hatch. Now both adults work assiduously bringing food to their young. Sixty days after hatching, in mid-February, the young have moulted their down and acquired their sea-going plumage. The whole process has taken just four months and has been completed before winter returns.

King penguins breed on the island too. They are much larger birds, standing a good nine inches taller than macaronis. That being so, their chicks need a longer time to grow to full size. Some adults arrive in November, but they take three weeks longer than the macaronis to incubate their eggs which are twice as big, and by the end of the summer, when the full-grown macaroni young are leaving, the young king penguins, although they have reached adult weight, are still are not strong enough to go to sea and are still clad in overcoats of thick furry brown down. They stand around in creches several hundred strong. Their parents come back regularly to feed them, each identifying its own offspring among the crowd by

Above: a king penguin colony, South Georgia

the sound of its call. But as summer moves into winter, the days darken, the temperature of the sea falls, and the shoals of the shrimp-like krill, the kings' dietary mainstay, begin to disperse. So food becomes much harder for the adults to find. Their visits to feed their young, perforce, become more infrequent and eventually cease. As winter sets in, the chicks start a long fast. Those that hatched late in the season – and some may have done so as late as mid-April – stand no chance of survival. Hundreds in the colony die.

The following spring the parents return and rejoin the survivors. Feeding is resumed and in the summer when the chicks are between fourteen and sixteen months old, they moult their down and swim away from the island. It is too late now for their parents to lay again, so they too leave the island and travel to their feeding grounds to regain their strength. As a consequence of this timetable, king penguins can only breed successfully once every two years.

Farther south, on the coast of the Antarctic continent, the summer is, of course, even shorter and the winter even more bitter. King penguins do not breed here, but their cousins, the emperors do. These birds weigh twice as much as kings. This may, in itself, be an adaptation against the cold. A body can only lose heat from its surface. The bigger an animal is, the smaller its surface in proportion to its volume. So a big body retains heat more efficiently than a smaller one of the same shape. But when it comes to breeding, this bigger size, together with an even shorter summer, presents the emperors with an even greater problem.

They solve it by adopting a strategy that involves the most extreme hardship endured by any warm-blooded animal – mammal or bird. Instead of starting their breeding cycle at the beginning of the Antarctic summer, they do so at the very end. At this time, March or early April, the fringe of ice that surrounds the Antarctic continent like a white collar is at its narrowest. The emperors land on its edge, shooting out of the sea like rockets, and then travel south across it, sometimes waddling upright, sometimes lying on their bellies and pushing themselves forward with their flippers, towards the land which is still likely to be several miles away. They never reach it. After several miles, they come to a patch of permanent ice that is likely to have served them as a breeding ground for many years. As many as 25,000 may gather in one enormous crowd and here they being to court. The male stands still, drops his head on his chest, takes a deep breath and lets out a series of

trumpeting hoots. These calls, repeated many times over days, eventually attract a female. The pair face one another, point their beaks towards the sky and stand there motionless for several minutes. They have become a pair. Thereafter the two waddle around together until the moment for copulation arrives. One of them points its head downwards; the other does likewise. Then they mate.

By now winter is approaching fast. The temperature is falling rapidly and the ice around the continent is extending outwards by as much as two miles a day. In May or early June, the female produces one large egg. The pair have made no attempt to make a nest for there is nothing with which to make it – except snow. She cannot leave her egg on the ice, for it would freeze. Instead, she places it on the top of her feet. Within hours, the male walks up to her and the two stand facing one another, breast to breast. He endeavours to take the egg from her, prodding at it with his toes. Often she is reluctant to surrender it, but eventually she lets it slide on to the ice. Within seconds, the male gets his toes beneath it, juggles it up on to the top of his feet and covers it with a fold of his densely feathered abdomen. Producing the egg has taken a significant proportion of the female's bodily reserves. She needs urgently to replenish them and she heads back to the sea.

The male is now entirely responsible for incubating the egg. He stands with his companions occasionally shuffling forwards for a few yards. But there is nowhere to go. Penguins of other species become exceedingly quarrelsome when they are

Above: male emperor penguins each incubating an egg in the darkness of the Antarctic winter

Right: adult emperor penguins with their downy-coated chicks

incubating but the male emperors cannot afford to be. As the winter winds begin to blow, the days darken, the temperatures fall, and the emperors huddle closer and closer together. They use their tiny stump of a tail as the third leg of a tripod and rest on their heels, with their toes inclined upwards keeping their precious eggs off the ice and snug within the brood pouch, where it is 80 degrees C warmer than outside. The blizzards increase in severity, the wind screams across the ice at 100 miles an hour and the males huddle still closer, their beaks drawn down to their chests so that the napes of their necks, pressed tightly together form a feathered roof with scarcely a gap between them. Those on the side facing the wind take the full brunt of its force. But not for long. They shuffle around the sides of the huddle and shelter in its lee, so there is a constant movement. They have nothing to eat. Midwinter comes and for a month there is total darkness, except for the shifting veils and curtains of the Southern Lights playing overhead.

After sixty days, the eggs hatch. The males, now close to starvation, manage to produce a little milky secretion from their gullets for their chicks. And at this critical moment, the females reappear. They have been travelling across the ice for a long time for since they came here to lay at the end of summer, the ice fringe around the continent has extended very considerably. Some may have had to walk a hundred miles. They exchange calls with their mates, the two recognising the sounds they learned during courtship, even after their three months of separation. The

females have full crops and bend down to regurgitate their little chick's first real meal. One might think that the males would be only too anxious to leave, but they seem reluctant to relinquish their young and most continue to brood for about ten days. Then they start the long trek back to the sea to take their first meal for almost four months.

Three or four weeks later, they are back and take over care of the chicks, allowing the females to return to the sea. The chicks at this early stage in their lives have not yet developed the ability to generate their own heat. If they were left unprotected on the ice, they would die within two minutes. Their parents take turns in fetching food and guarding them, but even so one in four chicks die within the first month. As winter slackens its grip and the sea ice begins to break up again, the journey from the breeding grounds to the sea gets shorter and the parents are able to increase the frequency of feeding. The chicks gather into their own huddles. By early November, they are beginning to lose their downy coats and a hundred and fifty days after hatching, they start to fledge. The parents now stop feeding them and all, in long processions, trail down to the sea and food.

No one knows how long it took for individuals of one branch of the ancestral king penguin family to grow a little in stature, change their nesting regime and so become emperors and succeed in colonising the ice fringe of Antarctica. Birds are, however, surprisingly swift to change their behaviour in order to extend their range and get the best out of new circumstances. A hundred and fifty years or so ago, a completely new kind of environment began to appear. Mankind started to build not only with relatively friendly stone or brick but with glass and metal. The streets between these new buildings were no longer covered with mud but sealed from the earth by layers of concrete and asphalt, and the vehicles that carried people and goods along them were not powered by horses, whose droppings had provided ample meals for many birds in the past, but by internal combustion engines which pumped out poisonous gases. Over vast areas, there was not a green leaf to be seen. The daily rhythm of light and dark was disrupted by blazing artificial lights that turned night into day. A more alien environment could hardly be imagined. Yet birds colonised it almost immediately.

In spite of the fact that virtually nothing edible grows in these new-style cities, they contain abundant food, for their human inhabitants are very wasteful. This refuse is not allowed to accumulate around their dwellings. Fleets of lorries tour the streets, collecting it and taking it away to dumps beyond the cities' outskirts. There great earth-moving machines roar back and forth, levelling the loads dumped by continuous processions of lorries. The air is heavy with the stench of decay. Tatters of plastic wrapping, which neither rot nor fragment, blow about in drifts and hang from snags and spikes like travesties of leaves. Smoke swirls from innumerable small fires. But mixed with the poisonous chemicals, the broken glass and the

battered ruins of discarded domestic machinery, there are the decaying remains of humanity's meals.

Only a few birds have the behavioural robustness to tolerate such conditions or the digestive systems to cope with the kind of diet to be found here. But for those that have, there is a never-ending supply of food. Just as flamingos form huge flocks on salt lakes because few other kinds of birds can accept such circumstances, so the one or two species of birds that have learned how to make a living on rubbish tips also gather there in spectacular numbers – seagulls in Europe, vultures in South America, kites in India and marabou storks in Africa.

Other species have moved into the city centres to live among the buildings themselves. Pigeons are among the boldest and most numerous. Their ancestors, wild rock pigeons, lived on the cliffs and in the caves of the coast. There they nested alongside kittiwakes and shags, but right from the earliest times, human beings welcomed them into their towns. Pigeons, whether as adults, squabs or eggs, were good eating and the birds managed to feed themselves by gathering scraps that were far too small for people to eat. Furthermore, the birds could provide fresh meat during the winter when the lack of winter foodstuffs meant that the only domesticated animals that could be left unslaughtered at the end of the summer were

Above: black-headed gulls on a refuse dump, Japan

breeding stock. The Egyptians and the Romans accordingly provided pigeons with special accommodation, tall towers fitted internally with ledges on which the birds could roost and nest. So the pigeon became one of the first species of bird to be domesticated by humanity.

Strains of them were developed for particular purposes, some specially plump for food, some to race or carry messages by exploiting the pigeons' ability to return to their nest sites no matter where they were taken, and some simply to delight their owners' eyes and fancies by becoming oddly coloured or physically deformed. These new strains inevitably escaped and interbred with the wild birds. So eventually, the modern town pigeons appeared, whose variable colours reflect their mixed ancestry. They build their untidy nests on ledges and balconies, in gutters and beneath eaves, just as their ancestors did on the cliffs of the coast. Some have so little fear of human beings that they will sit on the heads, shoulders or hands of anyone who offers them food. A few have even learned how to exploit urban transport systems. In London it is common enough to see them hop into the carriage of a tube train to prospect for crumbs on the floor inside, and hop out again just as soon as they hear the hiss of air that precedes the closing of the doors. A few brave individuals have even been recorded as staying inside as passengers and travelling to the next station before alighting.

Cities offer other kinds of food as well as refuse. Kestrels hover, head down, searching for mice that might be foraging among the refuse bins. The peregrine

falcon that hunted the wild rock pigeons of the coast, has followed its prey. Some use the spires of European cathedrals as look-out posts from which to select their victims, and in New York, they dive down into the canyons between the skyscrapers to pounce on their prey.

In many cities, the bright lights that shine throughout the night attract dense clouds of moths, crickets and other insects. Alpine swifts which have built their nests on buildings in the Swiss city of Geneva have changed their habits to become active after dark for then there are insect swarms which are denser than any they might find during the day.

A male eagle-owl, reared during a breeding programme in Stockholm Zoo and known as Karl-Edvard, became a city dweller by choice. He was released, fitted with a small radio transmitter so that his movements could be checked. Instead of flying off to the countryside, he established himself in the city centre and there met a wild female. The couple paired and on at least one occasion mated on the roof of the Grand Central Station. They built their nest inside the windowless ruin of a coal-fired power station and there raised a family. Karl-Edvard fed his chicks with rats which he caught on a nearby recreation field and regularly paid courteous visits to his caged parents in the zoo, sitting outside their cage and chatting. He also benefitted from the city's veterinary services for once, when he was greatly weakened by pneumonia, he was collected and successfully treated with penicillin. Eventually he moulted the tail feather that carried his transmitter and left it on the roof of the Cavalry headquarters. Thereafter he was able to conduct his affairs in privacy.

Perhaps the most ingenious exploiters of urban conditions are carrion crows that live in a Japanese city. The species is abundant in the mountains and forests of Japan but has also moved into urban areas. In 1990 the birds living in Sendai City somehow discovered that the green globes hanging from walnut trees that line some of the streets contained tasty nuts. But even though their beaks are hefty for crows, they were unable to crack the nuts for themselves. Nor did nuts break when dropped from the air, a technique used by the birds with other shelled morsels. The traffic along the town's streets provided them with a solution. Some of the birds wait beside the traffic lights at one of the cross-roads holding a walnut in their beaks. As soon as the lights turn red, the birds fly down and place the nuts in front of the cars. The lights turn to green, the traffic rolls forward over the nuts, and when it turns red again, the crows hop down into the road and hastily pick up the fragments of the kernels before the traffic starts to move forward once more.

City life may have more subtle attractions for birds than food. In Britain, on autumn evenings, starlings gather in flocks on the outskirts of towns. The hundreds grow into thousands. Then the whole immense assembly takes to the sky for aerial manoeuvres. They swirl in great clouds that swell and taper, coalesce and divide

Left: a kestrel surveys a city for prey, England

with the birds flying so closely together and with such finely co-ordinated dives, banks and swoops that it seems that they are all reacting to some common choreographic instructions. The display may last for half an hour or so. As one flock descends on a building and lands on its ledges and parapets, so another arrives to share in the aerobatics until darkness falls, the sky empties and the entire building is laced and dotted with perching birds. It may be that the starlings spend the night in town because it is a degree or so warmer there than in the countryside. The buildings provide a shelter from the wind and, no matter how well insulated they may be internally, inevitably radiate a little heat. It could also be that these immense roosts serve as information centres. Birds that have fed well during the previous day will, the following morning, fly straight back to same site to resume their meals, and birds that fared less well and have no such motivation will follow them. But why all should perform an aerial ballet is still unexplained.

The fact is that just as country people all over the world seem impelled to migrate to towns, even though when they get there they can find neither homes nor jobs, so some birds also find towns irresistible. Nowhere is this more dramatically apparent and inexplicable that in the central Brazilian town of Manaus. The town stands on the banks of the Amazon and on its outskirts, it has a large oil refinery. Like all such installations, this is filled with deafening unceasing noise. Plumes of flame spurt from pipes as waste gas is burnt off. Rhythmic blasts of steam hiss out in great clouds and condense into drifts of tepid drizzle. Many of the pipes are so hot that they are painful to touch. Others shudder with the violence of the reactions that are going on within them. It is difficult to imagine an environment that is more radically different from the pristine rain forest that still stands a mile or so away across the muddy river on the opposite bank. Yet every night, tens of thousands of purple martins leave the forest and fly over to the refinery. For ten minutes or so blizzards of them swirl down from the darkening sky, dodging aerobatically through the maze of metal, to settle in neat rows along the pipes, rails and ladders. Then the invasion is over and the refinery is festooned with birds. In the hot humid climate of the Amazon, it can hardly be that they have come for warmth. Nor is it likely that there are fewer predators around the refinery than in the forest, for hawks are often circling overhead. Maybe like fish in shoals and antelope in herds, they are seeking the safety that comes from numbers.

Whatever the reason that draws birds to towns, city dwellers all over the world welcome them. Many of the purple martins which roost in the Manaus oil refinery migrate each spring to North America. There they become some of the most pampered of all free-flying birds. Once they found homes for themselves in tree holes. Now human beings provide them with special lodgings. The tradition, it is said, was started by native people, the Choctaw and Chickasaw Indians, who were glad to see the birds when they arrived each spring and hung out hollow gourds to serve as additional nesting places for them. Today that tradition has been hugely elaborated. Martin lovers construct special tenements, that may be so big and elaborate

that they can accommodate several hundred birds. Some are assemblies of gourds or specially made fibreglass globes. Others are more fanciful and are made in the shape of Siamese temples, railway coaches, or streets from a town in the old wild west. Many are mounted on masts fitted with ropes and pulleys to allow the nest boxes to be lowered so that the human landlord can inspect the inside of the nests to check on the progress of the occupants or the hygiene of the nest linings. They all, however, must have an entrance hole of just the right size – big enough to allow the martin to enter but small enough to prevent starlings from doing the same. Just inside most, there is a baffle to prevent an owl or a raccoon trying to reach inside. As many as half a million people along the eastern United States care for martins in this way, and with such success that today hardly any of the martins in this part of America nest in any other fashion. As a consequence, their numbers are now very much greater than they would be if they still had to find tree holes in which to bring up their families.

This affection for birds on the part of humanity has over the past few centuries led to major changes in the range and distribution of many species. Specimens imported from overseas to be kept caged in collections, public or private, have frequently escaped and found homes for themselves in the wild. In the seventeenth

century, King Charles II had his own collection of birds in London's St James' Park. Canada geese were among them. By the middle of the eighteenth century, these handsome foreigners were living in English wetlands alongside native mute swans and falling to the guns of wildfowlers. A few may have been migrants from the Arctic that, flying south in autumn, lost their way and crossed the Atlantic by mistake, but most, certainly, were escapees. The conditions in Britain and northern Europe suited them very well. Today Canada geese are an established element in the British avifauna and flourish in such numbers that, in the eyes of many land-owners and farmers, they are a major pest, devastating fields of newly sown wheat and fouling lawns and meadows with their abundant droppings. Those most hand-some of ducks, the mandarins, were from the mid-nineteenth century onwards de-liberately encouraged to establish themselves in Britain by bird-loving land-owners who delighted to see flocks of them on their lakes. It took some time before they found a permanent place for themselves in the wild but now colonies are estab-lished in many areas in southern England and are, apparently, safer there than in their native China. The latest, and perhaps most spectacular and surprising exotic species to settle in Britain is the ring-necked parakeet from Asia. It was for long popular as a cage bird, but some eventually escaped. Surprisingly perhaps, these brilliantly coloured birds that seem to belong in a tropical landscape, do not find English winters too bitter. Now there are several large colonies in southern Eng-land, and exotic jade-green strangers with red beaks and purple collars are joining blue tits and spotted woodpeckers in collecting peanuts from feeders in suburban gardens.

During the nineteenth century, Europeans who had emigrated and settled on other continents also began to import birds. They were not content with the unfa-miliar and often glamorously coloured birds that they found in their new surround-ings but yearned for European species. In the 1850's, European house sparrows were released in Brooklyn and a hundred other cities in the United States and Can-ada on the improbable grounds that they might help keep local insects under con-trol – in spite of the fact that, as anyone could see, sparrows are primarily seed-eaters. By the end of the century they were nesting in every state in the Union. European starlings were liberated in New York's Central Park. Within fifty years they had spread right across the continent and had reached California in the west, Alaska in the north and the Mexican border in the south.

European colonists in Australia and New Zealand were particularly enthusias-tic about importing animals and plants of all kinds to supplement what they re-garded as the commercially unsatisfactory and aesthetically unpleasing native fauna and flora. Every State in Australia formed its own Acclimatisation Society which set about putting things right. Prickly pear, a spiny succulent from South America which spread and destroyed great areas of potentially productive pasture,

Left: a ring-necked parakeet feeding in a London garden

was originally imported by the infant colony of New South Wales to provide food for the cochineal insect – which itself was imported to provide red dye for the uniforms of the military. Rabbits were introduced for their flesh and their fur, and when their numbers in the wild began to increase astronomically, mongooses from India were released to keep down the rabbits – upon which they had little effect. Cats were brought by settlers who liked to have them as household pets, and foxes by those who found their pleasure in putting on red coats and galloping off to chase them through the groves of eucalyptus.

Naturally enough, bird-loving members of these Societies also set about improving the Australian avifauna. Secretary birds from Africa were introduced in hope that they might reduce the numbers of native snakes – but apparently had little success and did not survive long. Starlings and sparrows were introduced in bulk, as they had been in the United States, to control insect pests. They flourished mightily but ate few insects. Other British birds were released so that, in the words of one Society's prospectus, ' they may be permanently established here and impart to our somewhat unmelodious hills and woods, the music and harmony of English country life'. So hundreds of song thrushes, blackbirds, linnets, robins, redpolls, skylarks, goldfinches, greenfinches, bullfinches, chaffinches, bramblings, and yellowhammers were released into the Australian bush. Some species failed to

Above: a feral cat devouring a cockatoo, Australia

establish themselves but others increased so greatly in numbers that in some areas they became the most abundant of all the local birds.

The indigenous birds, of course, were badly oppressed by those foreign invaders that settled successfully. Before this time, the only mammalian carnivores they had had to face were the relatively clumsy Tasmanian devils which were, in any case, primarily carrion feeders, and the small quolls which the English settlers called first, 'native polecats' and then more simply, 'native cats'. But the pet European cats that escaped into the wild were much more efficient bird hunters. Stoats and weasels that had been imported to control the rabbits also found local birds a much easier prey. Rats that no one had planned to introduce but which had nonetheless travelled in the holds of ships and clambered ashore along the mooring ropes, pillaged their eggs and slaughtered their chicks. And European birds took their food and occupied their nesting sites.

The damage was at its worst in New Zealand. The original absence of any land mammals whatever had allowed the native birds to lower their defences to such a degree that they took little trouble to nest in safe places and many had lost their power of flight. The carnage began some twelve hundred years ago when the first human beings arrived. They were Polynesians who had come in canoes from warmer islands away to the north. They found the flightless moas, some species grazing the meadows, others living in the forests and browsing among the trees. All were good to eat and the Polynesians hunted them with great efficiency. As the human population of the islands grew, the native forests were reduced, and the hunting of the giant birds became more intensive. They were snared and trapped, driven into swamps and clubbed to death. By three hundred years ago all of the moas were extinct. The arrival of the Europeans and the alien mammals and birds they brought with them greatly accelerated the devastation of the native fauna. Now it was not just those birds that were hunted for food that suffered. Now it was anything that could be caught by a cat, a rat or a stoat. The Europeans settlers hastened the process of clearing the native forests in order to make space for their sheep and cattle. Few of the local birds could withstand these new pressures. At least fifty per cent of New Zealand's species of native birds became extinct, eighteen of them within the last century and a half.

Flightless birds on other islands elsewhere in the world were equally vulnerable. The first animal exterminated by European man in historic times was one of them and was massacred so swiftly that its very name has become synonymous with extinction – the dodo. It lived on the island of Mauritius in the Indian Ocean. It was probably a member of the pigeon family that, like so many island birds, had become a flightless giant. Its island sanctuary was so remote that human beings did not discover it until the beginning of the sixteenth century. It was not until the seventeenth century that ships called at Mauritius with any frequency but then the

sailors who landed there were only too glad to get fresh meat and had no difficulty in clubbing the huge defenceless dodos. The last of them was slaughtered in 1681, less than two hundred years after men had first landed on the island.

A giant flightless seabird, the great auk, which stood two and a half feet high, once lived in the northern Atlantic. In the sea it was a powerful swimmer able to out-pace a rowing boat, but when it came ashore to lay its eggs, then it too was easily caught. And it too was edible. As men's mastery of these tempestuous northern seas became greater so they found every one of the islands on which the great auks nested and the huge birds became increasingly scarce. The last British great auk was caught on St Kilda, a lonely islet lying in the Atlantic west of the Outer Hebrides, and beaten to death in the belief that it was a witch. The last of all was killed in Iceland in 1844.

The invention of efficient and accurate hand guns brought even fully flighted birds within easy range of human hunters. The most numerous bird ever to have existed is thought to have been the passenger pigeon. Only two hundred years ago, the skies above the grasslands in the central United States were regularly darkened by immense flocks of them. One was estimated to contain two thousand million individuals and took three days to fly past. A century ago, people began to notice that passenger pigeons were not as abundant as they had been. The full cause of their decline is now uncertain. Deforestation as the land was brought into cultivation is now thought to be one of the causes of their decline, but intensive hunting was also undoubtedly a factor. No one worried very much about their shrinking numbers until the species was on the brink of extinction. By then it was too late. The last wild passenger pigeon was sighted in 1889 and the last surviving individual, a lonely captive female named Martha, died in Cincinnati Zoo in 1914.

By the middle of the twentieth century, human beings were at last becoming aware of the damage that they were inflicting upon the world's wildlife and the movement to conserve began to gather momentum. There was no shortage of bird species in urgent need of help. The ne-ne, a goose that evolved on the lava fields of Hawaii, had been reduced to about 35 wild individuals and 15 captives. A group of dedicated ornithologists, among them Sir Peter Scott, started a programme of intensive captive breeding. This was so successful that by 1960 it was possible to take groups back to Hawaii and release them in the wild.

In New Zealand, ornithologists anxiously surveyed the wreckage of their native avifauna and started to compile censuses of the survivors. In 1970 a scientist visited the Chatham Islands, a small archipelago 550 miles east of the main islands. This was known to be the only home of a unique species of New Zealand robin. The species on the mainland is dark grey and white. The Chatham Island species is entirely black. The scientist found that because of introduced predators and changes in the islands' vegetation, the black robins now survived only on a small rocky islet surrounded by steep cliffs called Little Mangere. There were only eighteen individuals and even they were hard pressed. The forest on which they relied for food

had been almost destroyed to create a landing-pad for helicopters serving ships fishing for crayfish. Even the remaining patches had been badly damaged by huge numbers of mutton birds which were digging so many burrows that the trees were tumbling over. As the forest continued to degrade, so there was less and less living space for the robins. By 1976 only seven of them were left and of these, only two were females. They were almost certainly the rarest birds in the world.

The New Zealand Wildlife Service decided that desperate measures had to be taken if the species was not to vanish altogether. A large patch of native forest was still flourishing on the neighbouring rather bigger island of Mangere. A team lead by Don Merton tackled the job of trapping the surviving robins, transporting them across the hundred yard wide strait to the bigger island and then taking them to the forest on its far side The operation which was almost as hazardous for the trappers as it was for the trapped, was managed, amazingly, without the loss of a single bird. One of the females bred that year and in the following seasons other chicks were raised. But by now several of the birds of the original transfer had died of old age. By 1978, only five black robins were left and there were still only two females. Some way had to be found to maximise the eggs produced by these two females before some sudden and unpredictable disaster overcame them or they too died of old age. Time was not on the team's side. It was known that, if a female black robin lost her first clutch, she would lay a second time. So in 1981, eggs were taken one by

Above: a Chatham Island robin,
New Zealand

one from a nest and put in the nests of Chatham Island warblers, a much more abundant species. The technique worked. The female laid a second time. There were failures and problems in rearing the chicks, but by 1985 the population had grown to 38. Today there are around 150 and the species, though still extremely rare on a world scale, seems to be out of immediate danger in the islands on which it evolved. When they analysed the history of the rescue, researchers realised that all today's population are the descendants of just one of the those two original females. She was known from the colour of the identifying band on her leg as Old Blue. She had lived for thirteen years, twice as long as expected for her species. The black robin's escape from extinction could not have been narrower.

Encouraged by this success, Merton and his group went on to try to help one of the most dramatic and astonishing of New Zealand's birds, the giant nocturnal flightless parrot, the kakapo. In 1960 it was believed by most authorities that the species had been lost, but in that year, signs of them were found in Fiordland, the most remote tract of New Zealand's wilderness in the south-west corner of South Island. Five of them were trapped and taken into a reserve for study, but they all proved to be males. Keeping track of the fortunes of the birds in the wild was extraordinarily difficult because the only practical way to reach the ridges and valleys where they lived was by helicopter. In 1974 a male and a smaller individual that was thought to be a female were trapped and sent to an island sanctuary to see if they could be persuaded to breed. But they showed no sign of doing so and when eventually they both died the putative female was found to have been a male. More expeditions discovered eighteen birds in the wild, but they were widely scattered and seemed out of contact with one another. And all were judged to be elderly males. Hopes for the survival of the kakapo seemed very faint indeed.

Then in 1977 a new population of about two hundred was found on Stewart Island, a fragment of land, fifty miles across, lying off the southernmost point of South Island. This seemed to be a more viable and vigorous population and contained some undoubted females. Hope was renewed. But in 1982 cats somehow reached the valleys where the birds lived. Remains were found of no less than sixteen kakapo which they had killed. Attempts were made to exterminate the cats around the kakapos' refuge but everyone realised that they could never be entirely eliminated from such a big island with an established human population. The Stewart Island kakapo also seemed to be heading for extinction.

Once again, radical measures were called for. Merton and his team were given permission to catch the entire Stewart Island population and move them to four small islands that had been totally cleared of alien predators. Sixty-one birds were caught, but to everyone's dismay, once again the majority were males. The birds settled down well in their new homes, but by now it had been discovered that the conditions kakapo require to breed are particularly demanding.

The males are polygamous. Each has his own territory within which he creates a number of shallow saucer-shaped depressions in which during the night he crouches, puffs himself up and makes booming calls. The females, attracted by the sound, select one male, mate with him beside his auditorium and then go away to lay their eggs and rear their chicks with no further help from him. But the birds have to build up huge bodily reserves that almost double their weight in order to be able to do this – the males so that they can devote much of their time during the nights visiting their booming-bowls and calling, the females in order to produce their eggs and feed the chicks, without help, for many months. Even in the most favourable environments, they cannot do so every season. In their original homes, the birds probably relied on the fruit of the podocarp trees that had dominated the lowland forests to bring them into breeding condition. In Fiordland and Stewart Island much of these forests had been felled and alien predators had driven the survivors from the small patches that remained. The birds in Fiordland had been found living high in the mountains where there was probably not enough good food to bring them into breeding condition anyway. This explained the fact that there were no young birds in the population. If this was the case, then the survivors in their new predator-free homes might manage to live for the length of their natural life, but they would never breed either, for the vegetation in their new environment was also likely to be inadequate. In a technical sense, therefore, the kakapo was probably extinct already.

The only solution was to provide supplementary food. That was started in 1993. By now only nineteen females were left. In 1995, fifteen nests were discovered but of the thirty-two eggs that were laid in them only thirteen were fertile and hatched. In 1997, however, the booms of the males were heard more frequently than ever before, and by January 1998 seven young kakapo had hatched successfully. So now there is a real chance that this remarkable bird may yet survive.

In the United States, species of birds continued to become endangered even after the extermination of the passenger pigeon as more and more wild country was claimed by human beings for their own purposes. The biggest of North American birds, the noble whooping crane, which stands five feet tall, was once a relatively common sight. But the draining of swamps on which it relied for food reduced its numbers year after year. By 1945 only sixteen individuals remained. Once again, at the very last moment, conservationists mounted a last ditch rescue. Reserves were created for them. Hunters, who regarded any wildfowl of any kind as fair game were warned of the impending catastrophe and told how to distinguish the whooping crane from smaller commoner relatives such as the sandhill crane, in the hope that they would spare them. Even so, the number of whoopers remained dangerously low. At this stage, George Archibald came to their rescue. He established a captive flock at his research centre in Wisconsin. Here he pioneered the technique

Right: whooping crane chicks following Kent Clegg
Overleaf: young whooping and sandhill cranes following Kent Clegg in his microlite

that was later to be used by Don Merton in New Zealand, of taking the first clutch from the birds and putting them in the nests of adults of a different kind. Sandhill cranes, a much more common species, proved to be admirable foster parents

Archibald raised eighty-four chicks using this technique. But then a new problem appeared. The chicks had become imprinted on their foster parents and clearly regarded themselves as sandhill cranes. Thus when they reached breeding age they would not respond to whooper courtship displays and would only accept a sandhill crane as a partner. A new technique had to be devised. Today, Archibald and his team place the eggs in incubators. A chick, when it hatches, finds itself in an indoor enclosure. A head, with the black streaks on either side of the beak and a black patch on the back of the head which are the diagnostic signs of a whooping crane, suddenly appears through a trap-door in one wall. It cranes down on its long neck, picks up a particle of food in its beak and offers it to the youngster. It is a sleeve puppet on the arm of one of the rearing team. For the next two weeks, while the chick is at its most impressionable, this is the only moving object it sees. Its human foster parent remains out of sight guiding the actions of the puppet by watching the operation through a one-way mirror. Even when the chick ventures out into a paddock, it is placed under the care, not of a sandhill crane but of a human helper wearing a whooping crane costume.

But one final problem remains. Whooping cranes migrate. They fly south to spend the winter in New Mexico and the Gulf coast of Texas, travelling in family parties, guided by the parent birds. If the hand-reared youngsters were released into the wild, as all concerned with whooping crane protection hope they will be, then

they could well respond to an in-built migrationary urge and set off south. But how would they find their way and – even more difficult – how could they discover the few reserves in the south where whooping cranes can be sure of spending the winter in safety?

That problem is being tackled by an Idaho farmer, Kent Clegg. He has a passion not only for birds but for aeroplanes. He too has reared whooping cranes but uses a different technique from that used by George Archibald. A group of them were hatched out together, Kent believing that if they are kept as a flock, there will be no problems about imprinting later in life. He also trained them from an early age to come to his calls and to follow him. As they approached adult size, he took them out for daily walks, leading them from a motorised farm buggy. Then he introduced them to his microlite aircraft. As he takes off, so he calls to them and they follow, flying in formation in a line stretching outward from the aircraft's wing, as they would do if their leader were an adult bird.

The refuge in New Mexico is some eight hundred miles away to the south. The two previous years, Kent had led groups of imprinted sandhill cranes down there, so he knew the route and the problems, and that the project was feasible. Another experienced microlite pilot would join him in the air, to try to fend off any attacks by eagles. A support party would travel by road with collapsible pens in which the birds could be kept out of danger during the night.

On October 13 the caravan prepared to leave. The flock was small – eight sandhill cranes and just four precious whoopers. All twelve carried small radio transmitters so that they could be traced if, somehow, they got lost. Kent took off in his microlite and the flock followed him into the air, but as he headed south they wheeled round north and settled back on one of their familiar home fields. Kent tried again but without success. The young birds' tie with their home territory was too strong. So he loaded them on to a trailer and took them to another location, fifteen miles away, where they could not see the fields where they had grown up. This time all went well and Kent, with the twelve birds flying in a line from his wing-tip, headed south.

At mid-day, the birds landed for rest and refreshments. In the evening Kent led them off again and that day they covered ninety miles. On the third day, disaster struck. A golden eagle attacked. The bird was driven off by the escort plane, which fired special blank shells at it, but not before one of the whoopers had been struck out of the sky. Fortunately, the eagle had not been able to inflict the heavy body wound that is normally fatal, but had only gashed the whooper's thigh. The ground party managed to catch the bird and get veterinary help. The wound was sewn up, the patient was given a dose of antibiotics, and was transported for the next few days by road in a trailer. On the seventh day out, the whole party was almost totally scattered, no doubt quite unknowingly, by a squadron of jet fighters which roared over them at low altitude. The cranes, as always when frightened in the air, clustered round Kent's microlite – coming so close that they nearly collided with him. But on the ninth day, he led them down to the Bosque Del Apache Reserve near the Rio Grande. For two days, Kent stayed with them, leading the birds down to the swamp to mingle with the resident population of sandhill cranes. Then he left them.

Watching Kent Clegg patiently instructing his charges in how to find their way through the skies was both a touching and a paradoxical sight. Birds, after all, have been flying from continent to continent for at least a hundred million years and we have only just started to do so. But that is only one particularly vivid demonstration of how recently we have acquired our dominance. They, long ago, took up residence in the Antarctic, the coldest place on earth on which we only set foot less than two centuries ago. They live and breed in deserts where no human being has ever survived unaided for any length of time. They can stay in the air for a year or more and girdle the earth, which we have only learned to do within the last few decades.

We are now the most widespread competitors that birds have ever had to face. We are also by far the most powerful. We have already exterminated whole species of them by direct attack, but the greatest destruction we have wrought has been inadvertent – a consequence of the wholesale changes we have made to the face of the earth. That damage need not continue. We now have the knowledge and the skill to maintain all the wonderfully rich range of birds that still exists on earth in all its complexity and glory. All we need, as the new masters of this planet, is the will to do so.

ACKNOWLEDGMENTS

My primary debt in writing this book lies with the multitude of observers and researchers, twitchers, banders and listers, painstaking recorders and measurers, professional scientists, amateur naturalists and straightforward masochistic, anoraked, binocular-hung bird-watchers upon whose multitudinous findings the preceding pages are based. A list of them all would occupy far more space than is available here and in any case would be impossible to compile for no one knows all their names. Their skills, however, are extraordinary and subtle. The ability to watch for long periods is very demanding; the ability to see is much rarer than one might suppose; and the insight needed to understand even rarer still. My admiration of their work is huge – as is my debt.

The task of surveying the huge quantities of information that such people have produced was tackled initially by Sharmila Choudhury, helped later by Adam White. It involved consulting a great range of journals, some of international standing and others published initially for members of small regional societies, as well as attending conferences and visiting scientists both in the field and in their laboratories. From the summaries they produced came the framework for both the chapters of this book and the television series with the same title. There then followed two and a half years of work in the field for film teams and for myself, marshalled and deployed by the leader of the whole project, Mike Salisbury. The images those trips produced were so kaleidoscopic and so rich that, in retrospect, it sometimes becomes difficult to be quite sure whether I first saw a particular behaviour with my own eyes in the field or on a screen watching the film as it came back from the twenty-three cameramen who worked on the series. Each of them will have debts to local ornithologists whose experience and knowledge enabled them to put their hides in the right place at the right time and gave them the privilege of focusing their lenses on birds doing extraordinary things.

I too have such personal debts of gratitude. Among those I should wish to thank are David Gibbs and Kris Tindige in Irian Jaya; Bill Black, Richard Holdaway, Peter Jenkins, Rod Morris and Philip Smith in New Zealand; Andrew Cockburn, Ted and Ann Secombe, Lindsay and Janice Smith in Australia; Jack Schick on Lord Howe Island; Chris Feare and Robbie Bresson in the Seychelles; Hiromichi Iwasaki in Japan; Gerard Ramsawak in Trinidad; Gian Basili and Carlos Bosque in Venezuela; Miguel Catelino, Mike Hopkins, Elizabeth Garlipp and David Miller in Brazil; Mingo Galussio in Argentina; James R. Hill III in Pennsylvania, Susanne Connor in Idaho, Ann Schnapf and Bonnie Ploger in Florida, Ann Harfenist in Washington State, George Archibald in Wisconsin and Kent Clegg in Idaho; Peter

Wellnhofer in Germany; Alan Kemp and Arnold Hooper in South Africa and Gert Erasmus in Namibia; Amotz Zahavi in Israel; Thoswan Devakul and Pilai Poonswad in Thailand; and John Madunich and Pepy Aravalo in the Galapagos. The completed text was read in part or in whole by Lars Svensson, Angela Milner and my director colleagues all of whom have striven valiantly to steer me from error.

The whole project, however, depended throughout on the work of a team centred upon the BBC's Natural History Unit in Bristol which included not only directors and cameramen but many others who played essential roles in the whole enterprise. Their names are listed below and I thank them all unreservedly.

Executive producer
Mike Salisbury

Producers
Miles Barton
Peter Bassett
Fergus Beeley
Nigel Marven

Assistant Producers
Ian Gray
Phil Hurrell
Joanna Sarsby

Researchers
Sharmila Choudhury
Adam White

Production Co-ordinators
Melissa Blandford
Anne Holmes
Di Williams

Production Secretary
Yvonne Webb

Unit Manager
Cynthia Connolly

Finance Assistant
Martin Whatley

Music
Ian Butcher
Steven Faux

Graphic Design
Mick Connaire

Sound Recordists
Trevor Gosling
Chris Watson

Film Editing
Tim Coope
Martin Elsbury
Andrew Mort
Jo Payne
Vincent Wright

Dubbing Editors
Paul Fisher
Angela Groves
Lucy Rutherford

Dubbing Mixers
Martyn Harries
Peter Hicks

Photography
Andrew Anderson
Barrie Britton
Jim Clare
Lindsay Cupper
Stephen de Vere
Trevor de Kock
Justine Evans
Richard Ganniclifft
Geoff Gartside
Nick Gordon
John Hadfield
Nick Hayward
Richard Kirby
Mike Lemmon
Michael Male
Hugh Maynard
Ian McCarthy
Mark Payne-Gill
Mike Potts
Martin Saunders
Martin H. Smith
Gavin Thurston
John Waters

SOURCES OF PHOTOGRAPHS

Frontispiece, Aquila (Hanne & Jens Eriksen); **10** Bruce Coleman (Alain Compost); **13** Bruce Coleman (Kim Taylor); **15** Ardea (Francois Gohier); **17** Aquila (J.J. Brooks); **20** Planet Earth Pictures (Jon & Alison Moran); **22** National Geographic Image Collection (Louis Mazzatenta); **25** Bruce Coleman (C.B. & D.W. Frith); **26** Planet Earth Pictures (Thomas Wiewandt); **27** *above* FLPA (T. & P. Gardner), *below* Planet Earth Pictures (John Waters); **28** Ardea (Peter Steyn); **29** Bios (Seitre); **30** Hedgehog House (Tui de Roy); **33** Ardea (Pete & Judy Morrin); **35** Gerald Cubitt; **36** Ardea (Don Hudden); **37** Gerald Cubitt; **40** Masaaki Fujimoto; **41** Bruce Coleman (Dieter & Mary Plage); **42** Okapia (Hans Schweiger); **43** Bruce Coleman (S. Nielsen); **45** Bruce Coleman (Kim Taylor); **46** Windrush (Arthur Morris); **47** Animals Animals (Richard Day); **48** FLPA (Silvestris); **49** Oxford Scientific Films (Daniel J. Cox); **50** NHPA (Stephen Dalton); **51** Oxford Scientific Films (Stan Osolinski); **52** *above* NHPA (Stephen Dalton), *below* FLPA (Roger Wilmshurst); **53** Dave Richards; **54** Oxford Scientific Films (John Netherton); **55** Hedgehog House (Stefan Lundgren); **56** Planet Earth Pictures (R.L. Matthews); **57** Windrush (Chris Schenk); **58** *above* Auscape (Joe McDonald), *below* natural Science Photos (Carol Farneti Foster); **61** Oxford Scientific Films (C.M. Perrins); **62** FLPA (S. Maslowski); **63** Ardea (C. & J. Knights); **64-5** Windrush (Arthur Morris); **66** Okapia (U. Walz); **68** Oxford Scientific Films (Owen Newman); **70** Ardea (Robert T. Smith); **72** *left* Oxford Scientific Films (Neil Benvie), *right* Bios (Dominique Delfino); **73** Planet Earth Pictures (Mike Read); **74** Ardea (Edgar T. Jones); **75** Bios (Bengt Lundberg); **76** Bruce Coleman (Jeff Foott); **77** Ardea (J.B. & S. Bottomley); **79** NHPA (Kevin Schafer); **80** FLPA (S. Maslowski); **81** Ardea (Francois Gohier); **83** Luiz Claudio Marigo; **85** Oxford Scientific Films (Robert A. Tyrrell); **86** Bios (Seitre); **87** Planet Earth Pictures (Carol Farneti); **89** Animals Animals (Don Enger); **91** Oxford Scientific Films (Tui de Roy); **93** BBC Natural History Unit (Rico & Ruiz); **94-5** Bruce Coleman (Günter Ziesler); **96** Auscape (J. Ferrero-Labat); **99** NHPA (Alan Williams); **101** Oxford Scientific Films (Stan Osolinski); **103** Aquila (J.J. Brooks); **104** Oxford Scientific Films (Tui de Roy); **105** NHPA (Stephen Dalton); **106** Auscape (Nicholas Birks); **108** *above* Ardea (Clem Haagner), *below* Bios (J.C. Malausa); **109** Ardea (Ian Beames); **110** Planet Earth Pictures (Annie Price); **111** Ardea (S. Roberts); **112** FLPA (David Hosking); **114** *above* Bios (Alain Guillemont), *below* Günter Ziesler; **115** FLPA (Fritz Polking); **116** Bios (Mathieu Laboureur); **117** Oxford Scientific Films (Jorge Sierra); **119** Planet Earth Pictures (Paulo de Oliveira); **120** Natural Science Photos (L. Rubin); **121** Ardea (G. Threlfo); **122** Ardea (G.K. Brown); **123** Dave Richards; **124** Oxford Scientific Films (Tom Leach); **125** Okapia (Robert Gross); **126** Auscape (Jeff Foott); **127** Oxford Scientific Films (John Netherton); **128** Oxford Scientific Films (Mike Birkhead); **129** Oxford Scientific Films (Richard & Julia Kemp); **130** BBC Natural History Unit (Adam White); **131** Planet Earth Pictures (Sean Avery); **132-3** Bios (M. Denis-Huot); **134** BBC Natural History Unit (Hans Christoph Kappel); **135** Aquila (Kevin Carson); **136** ENP (Gerry Ellis); **137** Aquila (David Owen); **139** Hedgehog House (Tui de Roy); **140** Bios (Tony Crocetta); **141** Auscape (Tui de Roy); **142** *both* Natural Science Photos (D.B. Lewis); **143** Oxford Scientific Films (Doug Allan); **144** ENP (Konrad Wothe); **145** Hedgehog House (Tui de Roy); **146-7** NHPA (B. & C. Alexander); **149** Bruce Coleman (Dr. P. Evans); **150** Bios (Thierry Thomas); **152** Mike Salisbury; **155** Bruce Coleman (Konrad Wothe); **156-7** Premaphotos Wildlife (K.G. Preston-

Mafham); **158** Günter Ziesler; **159** Natural Science Photos (Amrit Pal); **160** Bruce Coleman (Dr. Scott Nielsen); **161** Bruce Coleman (Peter Davey); **162** BBC Natural History Unit (Tony Heald); **163** Animals Animals (Michey Gibson); **165** Planet Earth Pictures (Richard Matthews); **166** Nature Production (Goichi Wada); **167** Okapia (Dietmar Nill); **168** Bruce Coleman (Erwin & Peggy Bauer); **170** *above left* Planet Earth Pictures (William S. Paton), *above right* Planet Earth Pictures (Peter & Tristan Millen), *below left* Oxford Scientific Films (Carlos Sanchez), *below right* Bios (F. Cahez); **172** FLPA (D. Maslowski); **173** Günter Ziesler; **174** ANT/NHPA (Klaus Uhlenhut); **175** BBC Natural History Unit (Nick Gordon); **177** Bios (Michel Rauch); **179** FLPA (E. & D. Hosking); **181** Oxford Scientific Films (Roland Mayr); **183** Okapia (Konrad Wothe); **184** Natural Science Photos (Anthony R. Dalton); **185** Bruce Coleman (Tero Niemi); **186** Günter Ziesler; **187** Oxford Scientific Films (Stanley Breeden); **188-9** Bruce Coleman (Orion Service & Trading Co.); **190** ENP (Konrad Wothe); **191** Günter Ziesler; **192** Windrush (C. Tyler); **193** Oxford Scientific Films (Tui de Roy); **194** Bruce Coleman (George McCarthy); **197** BBC Natural History Unit (Richard Kirby); **199** *both* BBC Natural History Unit (Richard Kirby); **200** Oxford Scientific Films (Michael Dick); **201** Planet Earth Pictures (Ford Kristo); **202-3** BBC Natural History Unit (Richard Kirby); **205** NHPA (Bruce Beehler); **207** BBC Natural History Unit (Barrie Britton); **208** Oxford Scientific Films (Tom McHugh); **209** Oxford Scientific Films (Kenneth W. Fink); **210** *above* BBC Natural History Unit (Hugh Maynard), *below* BBC Natural History Unit (Nick Gordon); **211** Oxford Scientific Films (Michael Fogden); **212** Auscape (Wayne Lawler); **215** BBC Natural History Unit (Barrie Britton); **217** Windrush (Michael Gore); **219** Windrush (Kevin Carson); **220** Naturfotografernas Bildbyrå (B. Helgesson); **221** FLPA (David Hosking); **222-3** Günter Ziesler; **224** Ardea (B.L. Sage); **225** Günter Ziesler; **227** NHPA (Morten Strange); **229** Michael Gore; **230** Bruce Coleman (Antonio Manzanares); **231** Bios (Dominique Halleux); **232** Günter Ziesler; **233** Günter Ziesler; **235** Oxford Scientific Films (Tom Leach); **237** *above* Planet Earth Pictures (Thomas Dressler), *below* Okapia (Dr. Herman Brehm); **238** Ardea (W.R. Taylor); **240** Oxford Scientific Films (James H. Robinson); **242** Auscape (Glen Threlfo); **244** Bruce Coleman (Günter Ziesler); **245** Bruce Coleman (S. Nielsen); **246** Windrush (George Reszeter); **247** Bios (Gilles Martin); **249** *above* Aquila (Hanne & Jens Eriksen), *below* Gerald Cubitt; **252** Okapia (Robert Maier); **254-5** Bruce Coleman (Paul van Gaalen); **256** Auscape (Glen Threlfo); **257** Vireo (R. & N. Bowers); **258** Günter Ziesler; **259** FLPA (Roger Wilmshurst); **260** Okapia (Fritz Polking); **261** Bruce Coleman (Alain Compost); **262** Wendy Shattil/Bob Rozinski; **263** Oxford Scientific Films (Dennis Green); **265** *above* BBC Natural History Unit (Barrie Britton), *below* Wendy Shattil/Bob Rozinski; **266** Okapia (Gisela Polking); **267** Aquila (Mike Wilkes); **268** Natural Science Photos (Anthony Mercieca); **269** Bruce Coleman (Kim Taylor); **270** NHPA (Manfred Danegger); **271** BBC Natural History Unit (Cindy Buxton); **272** BBC Natural History Unit (Miles Barton); **273** *above* Auscape (Roger Brown), *below* Auscape (C. Andrew Henley); **275** Oxford Scientific Films (Richard Day); **278** Ardea (Hans D. Dossenbach); **280-1** Heather Angel; **283** Ardea (L.H. Brown); **284** Oxford Scientific Films (Bruce Davidson); **285** Auscape (D. Parer & E. Parer Cook); **286-7** Ardea (Jean-Paul Ferrero); **288** Hedgehog House (Gerald L. Kooyman); **289** Auscape (Graham Robertson); **290** Oxford Scientific Films (Doug Allan); **291** ENP (Pete Oxford); **293** Nature Production (Toshiaki Ida); **294** NHPA (Michael Leach); **296** NHPA (Hellio & Van Ingen); **297** BBC Natural History Unit (Adam White); **298** David Attenborough; **300** Planet Earth Pictures (Jiri Lochman); **302** Ardea; **303** Bruce Coleman (Dr. Eckart Pott); **304** ANT/NHPA (Brian Chudleigh); **307** Scott MacButch; **308-9** Scott MacButch.

INDEX OF BIRDS

The names of the birds used in the text are here accompanied by their scientific names, which identify the bird more precisely than do the English names. Numerals in **bold** type indicate illustrations.